D1085254

STP 1365

Cobalt-Base Alloys for Biomedical Applications

John A. Disegi, Richard L. Kennedy, and Robert Pilliar, editors

ASTM Stock Number: STP1365

ASTM
100 Barr Harbor Drive
West Conshohocken, PA 19428-2959

Printed in the U.S.A.

Library of Congress Cataloging-in-Publication Data

Cobalt-base alloys for biomedical applications/John A. Disegi,
    Richard L. Kennedy, and Robert Pilliar, editors.
        p. cm.—(STP; 1365)
    "ASTM Stock Number: STP1365."
    Includes bibliographical references and index.
    ISBN 0-8031-2608-5
        1. Cobalt alloys. 2. Metals in medicine. I. Disegi, John A.,
    1943- . II. Kennedy, Richard L., 1940- . III. Pilliar, Robert,
    1939- . IV. ASTM special technical publication; 1365.

    R857.C63 C63 1999
    610'.28—dc21   99-040202

Photocopy Rights

Peer Review Policy

Each paper published in this volume was evaluated by two peer reviewers and at least one editor.
The authors addressed all of the reviewers' comments to the satisfaction of both the technical editor(s)
and the ASTM Committee on Publications.

To make technical information available as quickly as possible, the peer-reviewed papers in this
publication were prepared "camera-ready" as submitted by the authors.

The quality of the papers in this publication reflects not only the obvious efforts of the authors and
the technical editor(s), but also the work of the peer reviewers. In keeping with long standing publication
practices, ASTM maintains the anonymity of the peer reviewers. The ASTM Committee on Publications
acknowledges with appreciation their dedication and contribution of time and effort on behalf of ASTM.

Printed in Fredericksburg, VA
October 1999

Foreword

This publication, *Cobalt-Base Alloys for Biomedical Applications*, contains 17 papers presented at the symposium of the same name, held on November 3 and 4, 1998, in Norfolk, Virginia. The symposium was sponsored by ASTM Committee F-4 on Medical and Surgical Materials and Devices. John A. Disegi from Synthes (USA), West Chester, Pennsylvania, Richard L. Kennedy from Allvac, Monroe, North Carolina, and Robert Pilliar of the University of Toronto, Toronto, Ontario, Canada presided as symposium chairmen and are editors of the resulting publication.

The scope of the symposium was intended to cover topics that have emerged in recent years such as alloy design, processing variables, corrosion/fretting resistance, abrasion and wear characterization, implant surface modification, biological response, and clinical performance. Although cobalt-base alloys are used extensively for a variety of dental, orthopaedic, neurological, and cardiovascular applications, the major portion of the publication is focused on orthopaedic applications.

The editors would like to express their appreciation for the help provided by two of the session chairmen: John Medley, Ph.D., from the University of Waterloo and Joshua Jacobs, M.D., from Rush Medical College.

We would also like to express our thanks to the ASTM staff that helped make the symposium and publication possible, most notably: D. Fitzpatrick for her help with symposium planning and E. Gambetta for the handling of manuscript submission and review. We are also indebted to the many reviewers for their prompt and careful reviews.

John A. Disegi
Synthes (USA)
West Chester, PA 19380

Richard L. Kennedy
Allvac
Monroe, NC 28110

Robert Pilliar, Ph.D.
University of Toronto
Toronto, Ontario, Canada M5S

Contents

Overview

Cast cobalt-base alloys were originally proposed for surgical implants over 60 years ago. Improvements in investment casting technology and a better metallurgical understanding of the cast Co-Cr-Mo system provided the technical justification to consider this alloy type for a variety of biomedical applications. Co-26Cr-6Mo investment castings performed reasonably well, but microstructural features and mechanical properties were not ideal for many surgical implant designs. Alloy processing considerations suggested that wrought versions of the cast grade material could provide metallurgical refinements such as better compositional uniformity, a finer grain size, higher tensile strength, increased ductility, and improved fatigue strength. Pioneering development programs were established between specialty alloy producers and implant device manufacturers to develop wrought cobalt-base implant alloys with enhanced metallurgical properties. The alloy development projects were successfully completed, and the first wrought low carbon Co-26Cr-6Mo composition was introduced in the 1980s for total joint prostheses. Wrought alloy versions were eventually used for orthopaedic, dental, neurological, and cardiovascular implant devices. These cobalt-base alloys provided a good combination of mechanical properties, corrosion resistance, and biocompatibility. As implant designs became more complex and the clinical applications were expanded, it became apparent that certain material features should be optimized. Some topics that have emerged in recent years include alloy design, processing variables, corrosion/fretting resistance, abrasion and wear characterization, implant surface modification, biological response, and clinical performance.

The symposium was organized to establish a forum for the presentation of new research and technical information related to the material issues that have been identified. The symposium and publication were divided into four major categories. This included: (1) Alloy Design and Processing (2) Mechanical Properties (3) Wear Characterization, and (4) Clinical Experience.

Alloy Design and Processing

Three papers were presented in this section which covered new alloy design schemes and innovative processing methods. The first paper by Tandon focused on the use of metal injection molding to provide near-net shapes. This work reviewed the processing parameters required to provide consolidated shapes with controlled properties. This work represented the first published study to examine this technology for Co-26Cr-6Mo alloy. Berry et al. provided important manufacturing information related to the production of a wrought high carbon analysis. Thermomechanical processing studies were aimed at optimizing the metallurgical structure in order to provide well-defined mechanical properties and improved wear resistance. The last paper in this section by the group at the National Institute of Standards and Technology investigated the potential of a new amorphous Co-20P alloy for orthopaedic applications. The surface characteristics of the electrodeposited film included excellent corrosion resistance, high hardness, and suggested future possibilities for exploiting this coating technology for cobalt-based implants.

Mechanical Properties

Six papers in this section emphasized the effect of microstructure modifications and processing variables on the mechanical properties of Co-Cr-Mo alloys. The paper by Becker and Bolton investigated the use of powder metallurgy techniques to provide a material with controlled porosity. The presentation examined the influence of powder compaction pressures and sintering atmospheres. The use of this technology was considered ideal for the manufacture of shaped acetabular cups with unique properties. The work by Berlin et al. highlighted the importance of post processing on the mechanical properties of investment cast and wrought alloy versions. Post processing operations such as abrasive blasting had no effect on fatigue, but sintering of porous coatings and laser marking reduced the fatigue strength of investment cast and wrought alloys. The post sinter fatigue strength of low carbon wrought alloy was dramatically reduced and was lower than the hot isostatically pressed ASTM F 75 castings. The third paper in this section by Mishra et al. included extensive metallographic examination, tensile testing, and axial tension-tension fatigue testing to compare investment cast versus high-carbon-wrought compositions with porous coatings. They concluded that the decreased chemical segregation and finer grain size may have been responsible for the improved fatigue strength observed for the porous-coated wrought high-carbon analysis. The presentation by Wang et al. explained the use of a powder metallurgy process to improve the sintering behavior of a Co-Cr alloy. The as-sintered fatigue strength was increased by a factor of X2 because of oxide dispersion strengthening and retarded grain growth during sintering. The thermally stable alloy permits the use of higher forging temperatures and more complex hip stem designs. Lippard and Kennedy reviewed the manufacturing operations for the production of wrought bar product intended for a variety of biomedical applications. Important technical information was documented for primary melting, remelting, hot rolling, annealing, and cold-working processes utilized for commercially available Co-Cr-Mo compositions. The effects of thermomechanical processing on the microstructure and tensile properties was presented for wrought low-carbon and high-carbon ASTM F 1537 material. Rodriguez described fundamental research on the role of face-centered cubic (fcc) to hexagonal close packed (hcp) phase transformation during plastic deformation of Co-Cr-Mo compositions containing low- and high-carbon content. High-carbon content and slow cooling after thermal treatment inhibited the metastable fcc $\Rightarrow$ hcp phase transformation. In contrast, a fast cooling rate after solution annealing and a controlled grain size range promoted phase transformation during deformation. The strain-induced phase transformation predominated when the carbon content was <0.05%, while the size, morphology, and distribution of secondary carbide particles controlled the ductility and fracture behavior at higher carbon levels.

Wear Characterization

The first paper presented in this session by A. Wang et al. investigated the heads of Co-Cr-Mo hip implants using the scanning electron microscope. Surface examination of cast, wrought-low-carbon, and wrought-high-carbon heads before and after hip simulator testing indicated evidence of third-body wear. It was postulated that residual grinding stone material introduced during the implant manufacturing cycle might have been responsible for scratches observed on the articulating surfaces of 15 implants that were examined. The next paper by K. Wang et al. evaluated the wear characteristics of various Co-Cr alloy combinations when tested on a reciprocating wear machine. In this study, self-mated as-cast ASTM F 75 material demonstrated lower wear rates than as-cast plus heat-treated couples. The wear resistance of as-cast hip heads mated with as-cast acetabular cups was also shown to be superior to various combinations of wrought-low-carbon and wrought-high-carbon components. Hip simulator testing also confirmed that self-mated as-cast couples demonstrated wear trends that were comparable if not better than new generation wrought-high-carbon metal-on-metal components. St. John et al. investigated the wear properties of hip heads and cups fabricated from high and low carbon-wrought Co-26Cr-6Mo alloy. Weight loss data generated in a hip simulator using EDTA stabilized bovine calf serum was statistically equivalent for sets of paired components manufactured from two types of ASTM F 1537 alloys. Killar and associates evaluated the effect of counterpart selection on the wear rate and surface morphology of ultra high molecular weight polyethylene (UHMWPE). Subsurface changes and wear rates of polymer cups in contact with implant quality ASTM F 138 stainless steel were more pronounced than with wrought Co-Cr-Mo alloy counterparts. Pellman presented an overview of the use of physical vapor deposition (PVD) coatings such as titanium nitride (TiN), zirconium nitride (ZrN), and diamond-like carbon (DLC) for medical devices. Wear, corrosion, and biocompatibility information was documented for these PVD films. Orthopaedic and dental applications were highlighted in addition to next generation coatings that are currently under development. Flores-Valdes et al. investigated a quaternary AlSi-FeMn intermetallic coating to improve the corrosion resistance and wear rate of cast Co-Cr-Mo alloy. A series of coatings were formed by reacting elemental powders in the temperature range of 873 to 1123 K. Continuous films deposited on cast F 75 material exhibited high hardness (1000 HV) and good interfacial adhesion.

Clinical Experience

Campbell et al. described the cellular response observed for clinically retrieved metal-on-metal hip components with CoCrMo bearing surfaces. CoCr particles that originated from the wear-in phase were responsible for tissue darkening (metallosis) reactions and included macrophages filled with black metallic particles in the nanometer size range. The wear debris was not associated with granuloma formation or necrotic tissue, but the authors stated that the long-term biological effects of in vivo wear products are not well defined. Hallab et al. analyzed serum protein factions from patients with cobalt-base total joint arthroplasty components. The distribution of serum Cr and Co concentrations implied that specific metal-protein complexes were formed from the implant degradation products. The physiological and clinical significance of high metal serum content is unknown according to the researchers.

Significance and Future Work

Modified cast Co-Cr-Mo compositions with enhanced thermal properties and expanded capabilities to provide porous coatings with improved fatigue properties represent significant metallurgical advances. The ability to produce complex shapes based on powder metallurgy methods is a major advantage for medical device manufacturers. The influence of grain size, secondary phases, and interstitial levels on mechanical properties has been better defined for wrought low- and high-carbon alloys. Fundamental research into cobalt-based phase transformations has provided the opportunity to improve the thermomechanical processing response. Numerous studies have evaluated the effect of surface modifications that influence wear resistance, and test protocols have been established to characterize tribological properties. Clinical researchers have a better appreciation of the mechanisms responsible for osteolysis and other unfavorable cellular reactions associated with the generation of implant wear debris.

Additional studies are needed to fully understand how alloy-processing variables can be fine tuned to control important material attributes. Future challenges include the need to standardize wear testing methods in order to compare results generated by different research groups. The use of wear-resistant implant coatings must be carefully evaluated from the standpoint of third-body wear phenomena. Sophisticated analytical examination of retrieved implant devices and perioprosthetic tissue should remain a high priority to expand our understanding of the material and design factors that effect clinical performance.

John A. Disegi
Synthes (USA)
West Chester, PA 19380

Richard L. Kennedy
Allvac
Monroe, NC 28110

Robert Pilliar, Ph.D.
University of Toronto
Toronto, Ontario, Canada M5S 1A1

Alloy Design and Processing

Rajiv Tandon[1]

Net-Shaping of Co-Cr-Mo (F-75) via Metal Injection Molding

Reference: Tandon R., **"Net-Shaping of Co-Cr-Mo (F-75) via Metal Injection Molding,"** *Cobalt-Base Alloys for Biomedical Applications, ASTM STP 1365,* J. A. Disegi, R. L. Kennedy, and R. Pilliar, Eds., American Society for Testing and Materials, 1999.

Abstract: Metal injection molding (MIM) is a recognized processing route for net- and near-net-shape complex parts for use in medical, automotive, industrial, and consumer industries. The MIM process holds great potential for cost reduction of orthopaedic implant devices for applications such as femoral components, tibial bases, cable crimps, and tibial trays. This study discusses the effects of processing parameters on the liquid-phase sintering behavior of injection molded ASTM F-75. Tensile test specimens were molded, debound, and sintered using different atmospheres. The static mechanical properties of the sintered alloys were compared to the cast and cast/HIP ASTM F-75. The liquid-phase sintered/solution-annealed MIM F-75 exhibited yield and tensile strengths greater than 550 and 900 MPa, respectively, with an elongation of 17%, thus exceeding the minimum requirements of the ASTM cast F-75. The HIP'ed and heat treated MIM specimens exhibited yield and tensile strengths of 500 and 1000 MPa, respectively, with an elongation of 40%. The sintering atmosphere played a major role in determining the static mechanical properties of the alloy, which can be partly attributed to the final carbon content. The maximum as-sintered density achievable was 8.2 g/cm^3. Since porosity is detrimental to the fatigue resistance, the as-sintered specimens were containerless hot isostatically pressed to eliminate any residual porosity. Rectangular test specimens were also molded, from which samples were machined in accordance with the ASTM Standard Practice for Conducting Constant Amplitude Axial Fatigue Tests of Metallic Materials (E466) to determine the fatigue properties of smooth and notched specimens. The MIM specimens performed similarly to the cast F-75, indicating a viable application of the MIM technology for F-75 implants.

---

[1]Metallurgist, Phillips Powder Metal Molding, 422 Technology Drive East, Menomonie, WI 54751.

Keywords: metal injection molding, liquid-phase sintering, orthopaedic implants, MIM F-75

Introduction

Metal injection molding (MIM) has emerged as a net-shape manufaturing route for complex parts for use in the medical, automotive, electronics, firearms, and consumer markets. The main attributes that make MIM a competitive technology versus investment casting and machining is cost. The ability to manufacture very complex parts at volumes ranging from 10,000 to over 1,000,000 parts per year, with a dimensional precision of ± 0.3% (or even better in some cases), and at typical cost savings ranging from 20 to 50%, has evolved MIM into an established manufacturing process.

The MIM process comprises of 4 steps: feedstock preparation, molding, debinding, and sintering [1,2]. Fine metal powders, with a typical mean particle size of 20 μm are mixed with an organic binder into a pellet shape suitable for feeding into the molding machine. Green parts are molded using a process similar to plastic injection molding. The next step is to remove the organic binder in a debinding step using either a solvent or solvent/thermal or catalytic process. The final step involves a high temperature sintering process in which shrinkage occurs. Typical values of linear shrinkage range from 12 to 25% depending on the starting ratio of the amount of metal powder to binder. This shrinkage factor is incorporated into the tool during moldmaking.

The two most commonly used MIM materials for medical applications are the 316L and 17-4PH stainless steels. Together, these two alloys represent the workhorse materials for almost all medical MIM applications such as forceps, jaws, surgical blades, orthodontic brackets, and staplers. However, the area of orthopaedic implant devices using the Co-Cr-Mo (F-75) has been largely undeveloped using the MIM technology. Applications such as femoral components, tibial trays, and cable crimps, represent a challenge in terms of their size, material performance, and biocompatibility issues.

Conventional wax-polymer based MIM processes are generally limited in their debinding step to a cross-section thickness less than 6 mm. Other water-based and water-soluble binder systems have shown slight improvements over the wax-polymer system in terms of their ability to debind relatively thicker cross sections. With the development of advanced acetal-based polymer system which allows for rapid catalytic debinding, it has become possible to mold thicker parts, in some cases exceeding 12.5 mm cross section thickness, while providing for excellent rigidity and virtually eliminating the need for support fixtures during debinding. Therefore, it has become feasible to consider relatively large parts such as tibial trays and femoral components within the envelope of MIM technology.

Apart from tool design and molding, the sintering behavior of F-75 is of critical importance to achieving a high performance product. Some of the variables affecting the sintering response of F-75 are the starting particle size, chemistry, and the sintering atmosphere.

It should be noted that the objectives of high temperature sintering in a MIM process are to obtain a part with high sintered density (typically exceeding 95% of the theoretical) and a homogeneous microstructure. On the contrary, the objective during porous coating Co-Cr-Mo beads is to maximize the surface area for tissue ingrowth while providing adequate strength. Consequently, the starting particle size and sintering temperatures are different for the two applications. The particle size is also important for MIM as it affects the sintering kinetics and the packing characteristics; the latter influences the shrinkage.

Within the relatively broad ASTM F-75 chemistry specification it is important to note that minor variations in levels of carbon can lead to significantly different sintering response and, hence, affect the density and mechanical properties. From a sintering standpoint, process control is much easier for a material containing little or no carbon, versus one requiring carbon within a tight tolerance. However, it is known that carbon is required to impart strength in F-75, hence, process control is expected to be a challenge with respect to the sintering atmosphere.

There are no published results of static and dynamic properties of F-75 via the MIM process. Since the starting particle size in the MIM process is relatively small, it is expected that an injection molded F-75 should be able to preserve a fine microstructure and achieve properties intermediate between a cast and wrought F-75. The optimization of the sintering process and the subsequent hot isostatic processing conditions dictate the microstructural parameters such as the grain size. The biocompatibility issue for injection molded F-75 is also not addressed in any study, although this characteristic is expected to be similar to the cast and wrought F-75 if the chemistry requirements are met.

This study discusses the process development of F-75 using the metal injection molding technology. Both static and dynamic properties were evaluated. As a benchmark, the performance of MIM F-75 was compared to cast F-75 whose properties are well characterized.

Experimental Procedures

Several powder vendors were screened in the inital phase of the development program. Basic powder characterization including particle size measurement, shape, tap density, and preliminary sintering studies to study microstructure evolution were used to select the powder for subsequent development. The nominal chemistry of the starting powder in this study was Co-29%Cr-5.9%Mo-0.4%Ni-0.8%Si-0.8%Mn-0.4%Fe-<0.1%C.

The powder was mixed with a proprietary binder, extruded, and pelletized for molding. Two geometries were used for molding test bars as shown in Figure 1. Dog-bone shaped tensile test specimens were molded according to MPIF Method for Preparing and Evaluating Metal Injection Molded Debound and Sintered Tension Test Specimens (Standard 50) with a L/D ratio greater than 4. The second geometry was a rectangular bar with approximate green dimensions of 12.7 x 12.7 x 106 mm. The molded parts were debound to remove about 90% of the total binder in the first step. The remaining 10% binder was subsequently removed by thermal decomposition prior to the high temperature sintering step. The sintering temperature, time, and atmosphere were optimized based on the microstructure. The sintering temperature range studied varied between 1340 and 1380°C. Three different atmospheres: reducing, neutral, and a proprietary atmosphere, were used to determine the effect on the sintering behavior and mechanical properties.

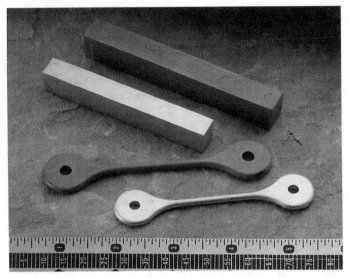

Figure 1 - Geometry of Test Specimens Used in the Study

Two different heat treating procedures were evaluated. The first heat-treat condition consisted of solution annealing the as-sintered specimen at 1250°C for 2 h followed by water quenching. It is well-reported in the literature that heat-treating a cast F-75 alloy in a temperature range from 1200 to 1250°C for 1 to 4 h results in ductility improvements due to dissolution of grain boundary carbides *[3,4,5]*. Therefore, the first heat treatment condition was used primarily to investigate the effect of porosity on the mechanical properties. The second heat treatment procedure involved a proprietary hot isostatic pressing cycle followed by solution annealing to eliminate any porosity.

The final dimensions of the rectangular bar after HIP'ing were approximately 10.5 x 10.5 x 90 mm, from which specimens for tensile and fatigue tests were machined in accordance to ASTM E8 and E466-96, respectively. The hour-glass shaped tensile bars had a gauge diameter of 6.35 mm with an L/D ratio of 5. The static mechanical properties were measured at a constant cross-head speed of 3.175 mm/min. The specimens for smooth axial tension-tension fatigue testing had a L/D ratio of 3.75 with a gauge diameter of 5.08 mm. The notched (Kt=3.0) specimens were machined to a 7.2 mm gauge diameter, with a gauge length of 19.81 mm. The fatigue tests were conducted at 60 Hz at R=0.1 and tested to 10 million cycles. The fatigue limit was determined as the stress level where 5 consecutive run-outs were obtained without failure at 10 million cycles. Note that all test specimens used for fatigue tests were HIP'ed and solution annealed to eliminate any effect of porosity.

Results and Discussion

The F-75 alloy exhibited a supersolidus sintering response in the range from 1350 to 1380°C. During supersolidus sintering, partial melting of the prealloyed powder occurs above its solidus temperature to form a liquid phase which provides for rapid densification via capillary force and particle fragmentation [6,7,8]. The sintering window for the F-75 alloy was confined to a narrow temperature range in the vicinity of 1370°C. The influence of the sintering atmosphere on the density and hardness is summarized in Table 1.

Table 1 - Effect of Sintering Atmosphere on the Density and Hardness of MIM F-75

Sintering Atmosphere	Sintered Density (g/cm³)	Apparent Hardness (HRC)
reducing	7.65 to 7.75	8 to 12
neutral	7.90 to 8.0	16 to 18
proprietary	7.95 to 8.20	20 to 25

For the samples sintered in the proprietary atmosphere, the apparent hardness was relatively insensitive to changes in the sintered density between 7.95 and 8.20 g/cm³. The samples sintered in a reducing atmosphere exhibited a much lower hardness between HRC 8 to 12 and a lower sintered density. The samples sintered in a neutral atmosphere showed properties intermediate between the above two. Note that the hardness reported represents the lower bound values by incorporating the effect of porosity.

The mechanical properties of the solution annealed, and the HIP/solution annealed bars processed using different sintering atmospheres are shown in Table 2.

Table 2 - Comparison of Mechanical Properties of MIM F-75 in Different Atmospheres

Sintering Atmosphere	Condition	Yield 0.2% (MPa)	UTS (MPa)	Elongation (%)	Reduction in Area (%)	Hardness (HRC)
Reducing	SA[1]	350	600	16	13	8
	HIP/SA[2]	300	900	25	22	18
Neutral	SA[1]	330	830	22	20	16
	HIP/SA[2]	270	950	28	25	15
Proprietary	SA[1]	550	925	17	15	25
	HIP/SA[2]	520	1000	40	25	25

[1] Solution annealed at 1250°C for 2 h followed by water quenching.
[2] Hot isostatically pressed followed by solution annealing.

The results show that the use of reducing and neutral atmospheres lead to relatively low yield strength as compared to the minimum requirements specified by ASTM F-75-92. The use of neutral atmosphere resulted in strength values that were intermediate between those obtained using the reducing and proprietary atmospheres. The best mechanical properties were obtained using a proprietary sintering atmosphere resulting in an outstanding combination of ultimate strength and ductility.

Solution annealing resulted in the elimination of grain boundary carbides while preserving the sintered porosity. Consequently, the ductility and ultimate strength improved over the as-sintered alloy, as expected. Hot isostatic pressing followed by solution annealing eliminated both the residual porosity and grain boundary carbides, but resulted in grain growth. Thus there was a corresponding decrease in the yield strength, but a significant increase in the ultimate tensile strength, reduction in area, and ductility. An important observation from Table 2 indicates the superior ductility of metal injection molded F-75 alloy as compared to its cast counterpart. Table 3 summarizes the mechanical property requirements as set by ASTM F-75-92, and compares the MIM values and typical values of cast F-75 used by implant manufacturers.

Table 3 - Comparison of Mechanical Properties of MIM F-75, Typical Cast F-75, and the Minimum Requirements of ASTM F-75-92

Material	Yield Strength (MPa)	UTS (MPa)	Elongation (%)	Reduction in Area (%)	Hardness (HRC)
MIM F-75	520	1000	40	25	25
Cast F-75 Typical	550	880	16	18	25-35
Cast F-75 Minimum	450	665	8	8	25-35

The ultimate tensile strength of MIM F-75 is higher by about 10 to 15% over typically observed values for cast/HIP/solution annealed F-75, and by almost 50% over the minimum requirement as set in ASTM F-75-92. Similarly, the ductility values for the MIM F-75 are significantly greater (by almost 100%) than the typically observed values for cast F-75. In fact, the mechanical properties of the as-sintered (proprietary atmosphere) and solution annealed MIM F75 as reported in Table 2 is superior to the minimum values listed in the above table. This indicates the viability of the MIM process in manufacturing Co-Cr-Mo parts for applications which do not require hot isostatic pressing to impart fatigue strength. Orthodontic brackets are a prime example of this type of application. The resulting improvements for the MIM F-75 are attributed to a more homogeneous microstructure as compared to the cast microstructure. Homogeneity during the MIM process is obtained using a combination of small starting powder size and optimizing the sintering parameters.

The fatigue endurance limit for the MIM F-75 alloy was determined to be 420 MPa using the smooth geometry which compares favorably with the cast F-75. Using the notched geometry, the fatigue endurance limit was about 210 MPa which is slightly lower than typical values of 240 to 260 MPa for cast F-75, despite the greater ductility of the MIM F-75 . At this point there is no explanation for the observed difference, which is a subject of ongoing study.

Two other issues, dimensional control and biocompatibility, have not been reported in this paper. Based on manufacturing experience, the dimensional control of components of similar relative size and geometry made using common MIM alloys like the 316L and 17-4PH range from 0.3 to 0.5%, or alternatively expressed as ± 0.076 mm/mm to 0.127 mm/mm [2]. The F-75 alloy is expected to show identical dimensional control to other common MIM alloys. The issue of biocompatibility is related to the material chemistry and is not expected to be of major concern as long as the chemistry specifications are met.

Conclusions

This is the first published study on the static and dynamic properties of F-75 processed using the metal injection molding technology. It is shown that the final properties are dictated by a strict combination of process variables including the sintering temperature, time, and atmosphere. The resulting static mechanical properties of the MIM F-75 are shown to be superior to the cast F-75. The dynamic fatigue properties of MIM F-75 also compare favorably with the cast F-75. The results of the study indicate the viability of the MIM process for manufacturing Co-Cr-Mo implants.

References

[1] German, R. M., and Bose, A., *Injection Molding of Metals and Ceramics*, Metal Powder Industries Federation, Princeton, New Jersey, 1997.

[2] *Powder Metallurgy Design Manual*, 2nd Edition, Metal Powder Industries Federation, Princeton, Jew Jersey, 1995.

[3] Mancha, H., Gomez, M., Castro, M., Mendez, M., and Juarez, J., "Effect of Heat Treatment on the Mechanical Properties of an As-Cast ASTM F-75 Implant Alloy," *Journal of Materials Synthesis and Processing*, Vol. 4, No. 4, 1996, pp. 217-226.

[4] Clemow, A. J. T., and Daniell, B. L., "Solution Treatment Behavior of Co-Cr-Mo Alloy," *Journal of Biomedical Materials Research*, Vol. 13, 1979, pp. 265-279.

[5] Dobbs, H. S., and Robertson, J. L. M., "Heat Treatment of Cast Co-Cr-Mo for Orthopaedic Implant Use," *Journal of Materials Science*, Vol. 18, 1983, pp. 391-401.

[6] Tandon, R., and German, R. M., "Particle Fragmentation During Supersolidus Sintering," *International Journal of Powder Metallurgy*, Vol. 33, 1997. pp. 54-60.

[7] German, R. M., "Liquid Phase Sintering of Prealloyed Powders," *Metallurgical and Materials Transactions*, Vol. 28A, 1997, pp. 1553-1567.

[8] Liu, Y., Tandon, R., and German, R. M., "Modeling of Supersolidus Liquid Phase Sintering, Part II: Densification," *Metallurgical and Materials Transactions*, Vol. 26A, 1995, pp. 2423-2430.

Graham Berry,[1] John D. Bolton,[2] John B. Brown,[3] and Sarah McQuaide[4]

The Production and Properties of Wrought High Carbon Co-Cr-Mo Alloys

---

Reference: Berry, G., Bolton, J. D., Brown J. B., and McQuaide, S., "**The Production and Properties of Wrought High Carbon Co-Cr-Mo Alloys,**" *Cobalt-Base Alloys for Biomedical Applications, ASTM STP 1365,* J.A. Disegi, R.L. Kennedy, R.Pilliar, Eds., American Society for Testing and Materials, West Conshohocken, PA, 1999.

Abstract: The medical implant market's requirement for a forged version of the normally cast only, high carbon (approximately 0.2%) variant of the Co-Cr-Mo alloy, which would combine increased wear resistance with the mechanical properties of the low carbon (approximately 0.05%) variant of Co-Cr-Mo, initiated an investigation to establish a viable manufacturing route for the high carbon variant. Initially, a single melted (vacuum induction melted) cast and forged route, and a metal spraying process were examined. Subsequently a vacuum induction melted, electroslag remelted and hot working route was developed using a number of compositional and thermo mechanical processing variants. The mechanical properties obtained on rolled and forged high carbon Co-Cr-Mo bar from 20mm to 50mm diameter were similar to those of the low carbon variant and also met the requirement of ASTM F1537-94 (warm worked). The wear resistance of the high carbon variant, measured using a pin on disc method, indicated some advantage over the low carbon variant at high applied loads.

Keywords: cobalt alloys (for surgical implants); mechanical properties; high carbon cobalt-chromium-molybdenum; implants; medical; forged; wear resistance

---

[1]Technical Manager, Firth Rixson Superalloys Ltd, Shepley Street, Glossop, Derbyshire. SK13 7SA UK

[2]Reader in Medical Engineering, Department of Mechanical & Medical Engineering, University of Bradford, West Yorkshire. BD7 1DP UK

[3]Metallurgist, Firth Rixson Superalloys Ltd, Shepley Street, Glossop, SK13 7SA UK

[4]International Exchange Student, California Polytechnic State University, College of Engineering, San Luis Obispo, CA USA.

Introduction

There have been a number of compositional variants of the well established implant alloy Co-Cr-Mo (nominally 28% Cr, 6% Mo, balance Co) and also a commensurate number of international, national and company specifications. These variants have often been associated with a number of alloy bar production methods for subsequent component forging or machining and for remelt stock for subsequent precision casting in ASTM F75-92 Cast Cobalt-Chromium-Molybdenum Alloy for Surgical Implant Applications . In general, a low carbon (approximately 0·05%) high nitrogen (approximately 0·15%) variant has often been used for the forging and machining applications, to the typical specification ASTM F1537-94, warm worked condition and a high carbon version (approximately 0·25%) has been used for casting application. In the latter case both high and low nitrogen contents have been manufactured.

The forged or machined low C, high N variants were generally used in those applications where high tensile, ductility and fatigue properties were required. The cast high C variants which were believed to be difficult to forge were generally used where wear resistance was more critical. The ductility and fatigue strength of the cast products were generally lower than those of the wrought low C, high N products.

The high C content was reported to improve the initial adherence of beads sintered on to stem implants, to aid fixation within the bone although bead detachment has been reported [1]. The sintering operation however, when applied to the low C high N material, required the use of higher temperatures which caused a reduction in the tensile strength and, accordingly, the fatigue strength. Thus, there became a demand from the market for a forged version of the high C Co-Cr-Mo which would provide not only improved wear resistance, particularly in metal to metal implants but also the facility for using lower bead sintering temperatures thereby minimising the reduction of fatigue properties.

The aim of this work has therefore been to develop a viable production scale process route for the manufacture of a forged version of the high C Co-Cr-Mo which would combine the potential wear resistance advantages with the mechanical properties of the warm worked condition of ASTM F1537-94.

In this investigation, a number of process routes and compositional variants of Co-Cr-Mo were examined before a technically, operationally and financially viable route was established. The development of both aspects together with the details of mechanical property and microstructural evaluations are described in this paper. A preliminary assessment of the wear resistance of the high C Co-Cr-Mo relative to the low C high N Co-Cr-Mo using the pin on disc method, is also reported.

Experimental Methodology

Since the target of the work was to establish a viable production route for forged high C Co-Cr-Mo, most trials were carried out on a production scale which involved cast quantities of up to 5000kg. Two parallel investigations were undertaken, one on process development and the other on compositional development, recognising that the two could be critically interdependent. Both investigations are reported.

Process Route Development

Initial Trials

The conventional <u>in-house</u> production method for the cast <u>high C</u> Co-Cr-Mo remelt bar stock has been to vacuum induction melt (V.I.M.) and cast into steel tube moulds of nominally, 75, 100 or 125mm diameter. After removal from the mould and surface preparation, the material is supplied to precision foundries for casting production. The initial trial was to attempt to forge 75mm diameter remelt bar stock (composition table 1.1) to about 30mm diameter thus giving a forging reduction (in area) of at least 50% which is generally considered to be sufficient to break down cast structures. Forging was carried out using an automated precision forging technique (GFM SX- 16 machine). However cracking problems were encountered and no suitable bar was obtained. Never-theless, for the purposes of comparison, the mechanical properties of test coupons (heat treated for 1 hour at 1220°C, air cool) cast from the remelt bar stock were obtained (table 2.1). As expected, the mechanical properties whilst meeting the requirements of ASTM F75, did not meet those of ASTM F1537 warm worked. The difficulty of directly forging remelt bar stock discouraged further development work on these lines and an alternative approach was attempted.

Using remelt bar stock from the same V.I.M. cast, a metal spray forming tech-nique (Osprey process) was used to produce three bars of approximately 180mm diame-ter. The process involved induction melting, transfer to a tundish and metal spraying techniques to build up a billet. The first bar produced was sprayed under a nitrogen at-mosphere and resulted in high nitrogen and oxygen contents of 0·55% and 0·35% respec-tively (table 1.2). Two further attempts, to reduce the gas content, were successful in meeting the specification nitrogen requirements of 0·35% maximum with levels of 0·056% and 0·059% (oxygen was 0·0070% and 0·0086% respectively).

Table 1.1 - *First VIM Trial, Cast C124, Cast Analysis, %*

C	Si	Mn	Co	Cr	Fe	Mo	Ni	N
0.210	0.73	0.33	65.80	27.26	0.19	5.48	0.18	0.008

Table 1.2 - *Compositions of Spray Material, Ingot Top Analysis, %*

Ingot No.	C	Si	Mn	Co	Cr	Fe	Mo	Ni	N
525	0.21	0.63	0.32	65.22	27.41	0.51	5.56	0.19	0.550
526	0.20	0.69	0.30	65.94	27.15	0.25	5.4	0.18	0.059
528	0.20	0.62	0.27	66.05	27.15	0.27	5.37	0.18	0.056

In the as-sprayed condition, the three bars contained porosity, so they were precision forged (1050°C to 1150°C) to 100mm diameter with the aim of improving the consolidation of the bars. A microstructural examination (unetched) revealed the presence of a surface layer containing oxidised, elongated porosity. The depth of this layer was from 3mm to 10mm depending on the gas content i.e. from 0·056% to 0·55% nitrogen, respectively. In each bar, there was also a heavy precipitation of pink grey particles (thought to be carbonitrides) at the grain boundaries. The precipitates were more elongated in the porous surface layer than in the centre and mid-radius positions. In addition to the above precipitates, the bars also contained a significant content of oxide particles. This was attributed to the process route used, where liquid metal contact with three refractory containers had occurred, i.e. during vacuum induction melting (VIM) and during the melting and pouring in the spraying operation. The microstructural examination of the etched structure revealed a fine grained structure of between ASTM 7 and 10 with some coarser grains (ASTM 5) in the centre of the bar. The grain boundaries contained carbides, in some cases in continuous films.

The examination generated concern about the material, namely the presence of porosity which could have been removed by grinding or machine turning, although this was a further operational stage resulting in significant yield loss. In addition, oxide cleanness was worse than that produced through the vacuum induction melting and electroslag remelting (E.S.R.) process. Thus the feasibility of the spray forming method proved to be unsuitable and it also had cost disadvantages over an "in-house" process route such as V.I.M. and E.S.R. This process, therefore, was not pursued further, though it did indicate that material with a finer initial microstructure was forgeable.

Having attempted two possible process routes for the production of wrought version of high carbon Co-Cr-Mo and found disadvantages to both, a further, "in-house" option was pursued. Both V.I.M. and E.S.R. facilities were available and were the predominant facility for the production of not only the low carbon (high nitrogen), forged Co-Cr-Mo but also a wide range of other superalloys. The spray forming trial had indicated that the high carbon Co-Cr-Mo could be forged if a finer initial "ingot" structure could be established. The use of an E.S.R. ingot provided an opportunity to examine these ideas further. Thus, a V.I.M. charge of 1250kg was melted and cast into a 230mm diameter electrode mould for electroslag remelting into a 305mm diameter ingot. This ingot was subsequently forged and rolled successfully to 30mm diameter bar. The details of this first trial and the subsequent evolution of the process and the compositions are described separately below together with the mechanical property and metallurgical assessments.

Further Process Route Development

The vacuum induction melting of the first trial cast of high carbon Co-Cr-Mo was carried out with a charge weight of 1250kg. Melting was carried out in a spinel bonded magnesia, alumina rammed lining. Virgin raw materials were used to produce a target chemistry of:

C 0.20%   Cr 28.5%   Mo 6.0%   Co Balance

Vacuum induction melting was carried out under a pressure of less than 20μbar and after refining for 30 minutes was cast into a 230mm diameter electrode mould. After cooling and stripping, the electrode was electroslag remelted at between 2·5 and 3·0 kg/min under a 70% CaF_2, 10% CaO, 10% MgO, 10% Al_2O_3 slag of depth 100mm using a hot slag start, collar mould of 305mm and retracting baseplate. The lower melt rate was used to minimise segregation effects and attempt to establish a finer grain structure suitable for forging. Subsequently the ingot was homogenised at 1150°C for 12 hours, again with the aim of minimising segregation effects and improving uniformity of carbide structure. Initially, there were concerns over the possibility of cracking or clinking so the initial ingot was furnace cooled after homogenisation.

The ingot was press forged to 90mm round cornered square (r.c.s.) from between 1050°C and 1150°C. The four billets produced were each cooled differently to assess the potential effect on clinking (air cool, coffin cool, stress relieve).

After minimal dressing, the four 90mm r.c.s. billets were cold cut in 3 and subsequently precision forged to 63mm diameter. Because no clinking was seen at the 90mm r.c.s. stage, the twelve 63mm diameter precision forged bars were air cooled.

The 63mm diameter billets were overall ground, and a) precision forged to 43mm diameter, b) rolled to 29mm diameter and directly hot straightened.

In order to simulate subsequent component forging operations, some of the 63mm diameter bar was hammer forged to 22mm diameter test coupons.

Process developments were subsequently aimed at addressing the following factors:

1)   The removal of precautions put in place to minimise or prevent clinking of electrode, ingot and billet.

The proportional and integral response of the programme controlling the E.S.R. operation was desensitised and the furnace cooling operations after homogenisation and forging were omitted because no clinking had been encountered.

2)   The need to modify the composition to ensure that the mechanical properties of the bar met the required specification level.

The nitrogen content was increased and this required not only a nitrided product to be added during the latter stages of V.I.M. but also required an increase in furnace pressure to retain it.

3)   The establishment of a more cost effective route including the scaling up to larger production units.

The weight of the "parent" V.I.M. cast was increased from 1250kg to 2500kg and then, using a larger V.I.M. furnace, to 6000kg, thereby producing 5 electrodes for E.S.R. The hot working of the ingot, billet and bar was also modified to use a larger intermediate section size. Otherwise, the process route remained as the initial trial.

Compositional Developments

A summary of the changes in composition is given in table 1.3 to 1.5. After the production and assessment of the first cast, two main changes were made. Firstly the nitrogen content was increased from 0·006% (no addition of nitrogen) to 0·13%, as a means of increasing the proof and tensile strength to the ASTM F1537 warm worked, level.

Secondly, some intermetallic (sigma) phase was observed in the first cast produced, so the ratio of (Cr+Mo) to Co was lowered to avoid this phase. Both changes were successful in achieving the targets though further increases in nitrogen content to 0·17% did result in some difficulties in cold straightening so nitrogen levels have reverted to 0·13%.

Table 1.3 - *First Production Cast, E6844, ESR Top Analysis, %*

ESR CastNo.	C	Si	Mn	Co	Cr	Fe	Mo	Ni	N
E6844	0.194	0.10	0.68	64.20	28.50	0.19	5.92	0.14	0.006

Table 1.4 - *Subsequent Production Casts, High Nitrogen Content, >0.14%, ESR Top Analysis, %*

	C	Si	Mn	Co	Cr	Fe	Mo	Ni	N
Mean	0.2030	0.181	0.675	65.387	27.552	0.293	5.501	0.242	0.160
Sigma	0.0069	0.138	0.016	0.548	0.404	0.052	0.175	0.076	0.011
Max.	0.216	0.50	0.71	65.75	28.54	0.41	5.69	0.38	0.176
Min.	0.190	0.08	0.65	64.04	27.2	0.24	5.19	0.12	0.141

Table 1.5 - *Subsequent Production Casts, 0.10 to 0.13% Nitrogen Content, ESR Top Analysis, %*

	C	Si	Mn	Co	Cr	Fe	Mo	Ni	N
Mean	0.2032	0.231	0.685	65.132	27.378	0.411	5.573	0.281	0.120
Sigma	0.0094	0.120	0.038	0.592	0.180	0.177	0.126	0.218	0.007
Max.	0.220	0.49	0.75	65.99	27.73	0.68	5.74	0.77	0.129
Min.	0.186	0.06	0.61	64.27	27.04	0.20	5.29	0.13	0.105

Mechanical Property and Metallographic Assessments

Mechanical Properties

The mechanical properties of the initial cast produced did not consistently achieve the target tensile strength levels though the simulated customer hammer forging operation not only enhanced the strengths to the target levels but also improved the ductility (table 2.2). Because the other results were below target, nitrogen levels were increased from ·006% to between 0·12% and 0·17% in subsequent casts. Reasonable correlations (R = 0.68 and 0.71) between proof and tensile strength levels and nitrogen content were established (figure 1) with scatter being attributed (tentatively) to test piece preparation and final hot working (rolling) reduction. Similar trends are also present in the low carbon, high nitrogen variant of Co-Cr-Mo. For the high carbon Co-Cr-Mo results (figure 1), it appears that the nitrogen content could be reduced below 0·12% but there would be

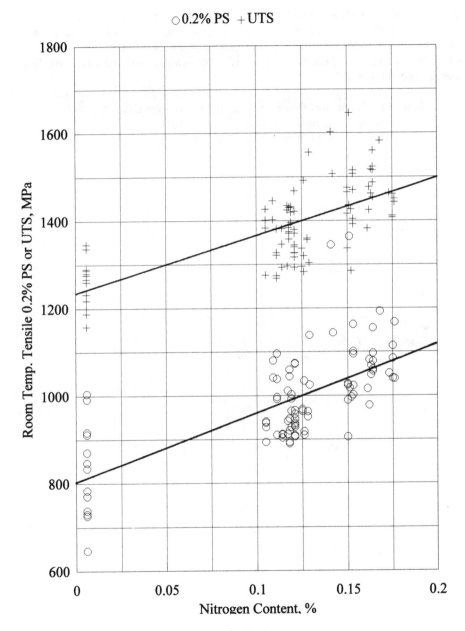

Figure 1 -
Room Temp. Tensile Strength vs Nitrogen Content

an increased risk of failure to meet the target properties. After production of over 30
E.S.R. casts, proof strength results have ranged from 891 MPa to 1364 MPa and tensile
strength from 1268 MPa to 1646 MPa. The target ductility level was also generally
achieved, though a limited number of results (five out of about one hundred) were below
specification. (Tables 2.3, 2.4).

Table 2.1 - *Mechanical Properties of Cast Samples from C124,*
       (Room Temperature Tensile)
       (Heat Treated 1 hour at 1220°C, Air Cool)

0.2% PS, MPa	UTS, MPa	Elongation, %
469	719	13.0
474	775	15.8
482	817	17.5
478	799	16.4
462	729	18.7
480	854	21.3

Table 2.2 - *Mechnical Properties, First Production Cast, E6844,*
       Room Temperature Tensile Results

	0.2% PS,MPa	UTS, MPa	Elongation, %	R of A, %
	726	1276	15	17
	646	1158	17	18
	871	1287	12	12
	738	1232	10	15
	730	1290	21	20
	784	1188	8	10
	771	1280	12	16
	847	1346	12	14
*	917	1218	20	23
*	834	1337	33	29
	912	1200	10	13
	992	1267	9	13
	1005	1260	10	14

*Hammer Forged

Table 2.3 - *Mechanical Properties, Subsequent Production Casts,*
>0.14% Nitrogen Content, (Room Temp.Tensile)

	0.2% PS, MPa	UTS, MPa	Elongation, %	R of A, %
Mean	1079.6	1470.2	17.9	18.4
Signa	94.9	66.2	6.7	5.6
Max.	1364	1646	38	38
Min.	906	1337	8	11

Table 2.4 - *Mechanical Properties, Subsequent Production Casts,*
0.10 to 0.13% Nitrogen Content, (Room Temp. Tensile)

	0.2% PS, MPa	UTS, MPa	Elongation, %	R of A, %
Mean	969.5	1368.5	17.3	18.1
Signa	62.7	62.4	4.4	3.6
Max.	1138	1556	26	27
Min.	891	1268	7	10

Microstructural Examination

Microstructural examination of mainly rolled bar and some precision forged bar have been carried out on all casts produced. Apart from the first cast produced where some sigma phase was observed, the microstructures (up to X500) of the rolled bar samples were similar. Thus in the unetched condition, little was observed though some primary carbonitrides (pink, grey) stringers and globules were present. Few, if any, oxide particles were observed. The typical cleanness assessment (ASTM E45) is given in table 3. This low level is expected as a result of the process route adopted i.e. V.I.M. plus E.S.R. Some occasional angular nitride particles were also observed.

Table 3 - *Cleanness Assessment, ASTM E45 Method D, Typical Result*

Severity Level Number	Type A Thin	Type A Heavy	Type B Thin	Type B Heavy	Type C Thin	Type C Heavy	Type D Thin	Type D Heavy
0.5	0	0	0	0	0	0	15	7
1.0	0	0	0	0	0	0	11	4
1.5	0	0	0	0	0	0	8	2
2.0	0	0	0	0	0	0	0	0
2.5	0	0	0	0	0	0	0	0

Average of Six Specimens

In the etched condition, a fine grain structure of ASTM 10 or finer was observed (figure 2) and there was some banding of the primary carbide stringers. Grain boundaries contained discontinuous carbides, reprecipitated following thermo mechanical processing (figure 2).

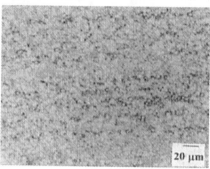

Figure 2- *Optical microstructure of high carbon Co-Cr-Mo alloy, mechanically polished and electrolytically etched in 5% aqueous HCl .*

Examination of the Relative Wear Rates of the High and Low Carbon Variants of Wrought Co-Cr-Mo

Test Procedures

Pin on disc wear testing was carried out on the high C and low C versions of Co-Cr-Mo, both meeting ASTM F1537 warm worked standard, and was conducted purely in an attempt to rank the wear performance of the two variants on a comparative basis. The testing procedures used were not capable of measuring the potential wear performance of either alloy when used under the conditions encountered as an artificial hip or knee joint replacement. The two alloys were tested as pins running against a disc made of high carbon grade material, the analyses of which are shown in Table 4.

Table 4 - *Compositions of Co-Cr-Mo alloys used for pin on disc wear tests*

Alloy	%C	%Co	%Cr	%Mo	%Mn	%Fe	%Ni	%Si	%N
High C Pin	0.2	Bal	27.4	5.45	0.68	0.4	0.25	0.15	0.16
Low C Pin	0.05	Bal	27.8	5.6	0.7	0.26	0.17	0.1	0.17
High C Disc	0.21	Bal	27	5.65	0.68	0.25	0.2	0.45	0.11

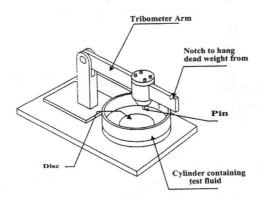

Figure 3- *Schematic of the pin on disc wear test rig*

Tests were conducted using the apparatus shown in Figure 3 which could be described as a disc driven by an electric motor at a constant speed of 210 rpm with the motor being mounted in such a way to ensure that vibrations did not affect the test. A stationary pin holder mounted onto a pivoted lever arm produced pin on disc contact with the pin at an inclined angle and with a load applied by dead weights hung from the end of the lever arm. Use of this inclined angle in combination with a rounded pin end, permitted several tests to be conducted on one sample simply by rotating the test specimen to a new position.

Both the pin and disc were fully immersed in a solution of aqueous 0.2 M potassium chloride (KCl) held at room temperature that was intended to provide an approximately *in vitro* corrosive environment during the wear test. The solution was contained within a Perspex cylinder which surrounded the disc assembly.

Wear pins, 38 mm long and 8 mm diameter, were machined from 28 mm diameter hot rolled bar, such that the pin axis was along the rolling direction. These were subsequently ground to form a rounded end of 5.5 mm radius that was used to form the wear test surface after manual polishing to a mirror surface finish with 1 micron diamond paste.

Disc specimens were prepared by slicing from a 50 mm diameter precision forged high carbon grade alloy such that their wear test surface was transverse to the forging direction. The test surface was polished to 1 micron diamond finish and their Ra surface roughness values were checked by Talysurf machine to ensure that they fell below the 0.05 μm maximum level specified in accordance with ISO 7206-2: 1996, Implants for Surgery - Partial & Total Hip Joint Prostheses, Part 2, Articulated Surfaces Made of Metallic, Ceramic, and Plastic Materials, International Standards Organisation.

Wear rates, expressed as weight loss per unit distance of travel, were assessed by measuring the weight loss that occurred in pins of both high and low carbon content by varying the circumference of the wear track formed on the disc, the duration of each test, and the load applied. Scanning Electron Microscopy (SEM) examinations of the wear scar were carried out to determine possible mechanisms of wear damage. Theoretical mass loss was also estimated by measuring the diameter of the wear scar from SEM micrographs, and by calculation of the volume lost from the relationship, [2]

$$V = \pi h^2(3r-h)/3 \qquad (1)$$

where

V = volume lost
r = radius of pin,
h = depth of wear scar.

Fairly high loads were initially applied to the wear pins to generate rapid wear and to provide a quick means of ranking the materials. Loads were later reduced to much lower levels in recognition of the fact that the spherical pin on flat plate geometry produced very high initial contact pressures that were well above those experienced under normal operating conditions. Initial contact pressures for each load are shown in Table 2 and were calculated for Hertzian contact between a ball and flat plate using the relationship, [3,4]

$$P = (6\ F\ E*^2\ /\pi\ R\ ^2)^{1/3} \qquad (2)$$

where

P = contact pressure
F = applied load
$1/E*$ = 1/Effective Young's Modulus = $(1-v_1^2/E_1) + (1-v_2^2/E_2)$
where E_1, E_2, v_1, and v_2 are the respective Young's Modulus and Poisson's Ratios of the disc and pin materials
R = pin radius

The flattened area of contact produced after wear had taken place significantly reduced these pressures and were estimated from the size of the wear scar as shown in Table 5.

Table 5 - *Hertzian contact stresses between the pin and disc at the start of the wear test plus estimated contact stress after wear of the pin*

Applied Load, N	Initial Contact pressure, MPa	Initial Contact shear stress, MPa	Final Contact Pressure, low carbon, 2 hrs, MPa	Final Contact Pressure, low carbon, 4 hrs, MPa	Final Contact Pressure, high carbon, 2 hrs, MPa	Final Contact Pressure, high carbon, 4 hrs, MPa
10.5	1,613	500	-	-	-	-
61.7	2,911	902	-	-	-	-
89.8	3,300	1,023	21.4	15.3	33.6	28.5

Results

Wear Rates

Measured wear rates for a number of tests on each material and expressed as weight loss per distance travelled and as weight loss per unit load per unit distance are shown in Table 6.

Table 6 - *Measured wear rates as a function of load and carbon content*

Normal load N	Wear rate μg/m High carbon	Specific wear rate μg/N/m High carbon	Wear rate μg/m Low carbon	Specific wear rate μg/N /m Low carbon
10.5	0.08	0.01	0.11	0.01
	0.05	0	0.08	0.01
	0.09	0.01	0.11	0.01
Average	*0.07*	*0.01*	*0.1*	*0.01*
Std deviation	*0.02*	*0*	*0.02*	*0*
61.7	-	-	0.16	0
	0.31	0.01	0.41	0.01
	0.21	0	0.37	0.01
	0.35	0.01	0.29	0
	0.35	0.01	0.26	0
	-		0.16	0
Average	*0.31*	*0.01*	*0.28*	*0*
Std deviation	*0.06*	*0*	*0.1*	*0*
89.8	0.85	0.01	0.6	0.01
	0.37	0	1.35	0.02
	0.21	0	0.86	0.01
	0.75	0.01	1.16	0.01
	-	-	0.71	0.01
Average	*0.55*	*0.01*	*0.94*	*0.01*
Std deviation	*0.26*	*0*	*0.28*	*0*

Examination of Wear Damage

Wear damage that occurred in the high carbon alloys at the very highest loads showed evidence of both adhesive and abrasive wear. Adhesive wear damage caused by adhesion between the pin and disc materials plus the presence of large grooves caused by third body adhesive wear were evident in several areas of the wear scar, see Figures 4 a & b.

Less evidence of severe adhesive wear was found in the low carbon grade alloys when tested at high loads but third body abrasive wear was still much in evidence, see Figure 5a. Corrosive pitting attack was detected in both the high and low carbon materials when tested at high loads but was much more in evidence for the low carbon grade alloy, see Figure 5b.

Adhesive wear was not detected to any significant extent in either the high or low carbon alloys when tested at the lowest loads (10.5 N) and wear surfaces were much smoother than those seen at high loads. Less grooving of the surface caused by ploughing of the surface by wear debris occurred in both the high and low carbon alloys, see figure 6a & b, but corrosion pitting attack was much more in evidence with the low carbon alloys, see Figure 6b. Evidence in support of pitting corrosion attack arose from energy dispersive analysis which detected small levels of chloride containing corrosion products within and around the pits.

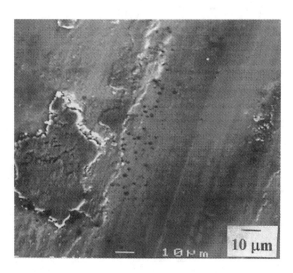

Figure 4a - *Adhesive wear damage to the surface of the high carbon grade alloy, tested at 89.9 N load. Note small area of corrosion pitting attack adjacent to adhesion zone. SEM Secondary Electron Image*

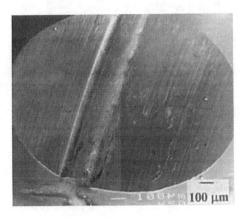

Figure 4b - *Third body adhesive wear caused by ploughing of the surfaceby wear debris particles. High carbon alloy tested at 89.9 N load. SEM Secondary Electron Image*

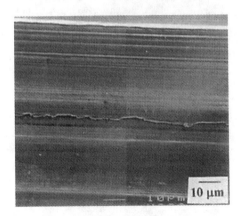

Figure 5a - *Low carbon alloy tested at 89.9 N load, mainly abrasive wear with evidence of plastic deformation of the tongues of material displaced by ploughing of the surface by third body wear. SEM Secondary Electron Image*

Figure 5b - *Corrosion pitting of the surface in a low carbon grade alloy tested at 89.9 N load. SEM Secondary Electron Image.*

Figure 6a - *Third body abrasive wear in a high carbon alloy tested at low load (10.5). SEM Secondary Electron Image.*

Figure 6b - *Third body abrasive wear in a low carbon tested at low load (10.5N). Extensive corrosion pitting of the surface by corrosive attack atthe base of wear grooves. SEM Secondary Electron Image*

Hardness Tests

Hardness tests carried on the polished surface of both high and low carbon grade alloys revealed some interesting effects that appeared to be caused by the actions of a mechanical abrasive polishing process.

Bulk hardness values measured on a polished surface prepared by conventional metallographic wet grinding and diamond polishing methods and by using the Vickers Diamond Pyramid Indentation method at 10 kg load proved that the high carbon grade alloy was significantly harder than the equivalent low carbon grade alloy. Somewhat different results appeared however when hardness of the surface was measured by micro hardness at a small load of only 30 g, both in the as polished condition and after electrolytically etching the surface in 5% aqueous HCl to remove the mechanically polished surface layer. These results shown in Table 7 suggest that mechanical polishing caused significant hardening of the surface in the low carbon alloy but had little effect on the surface hardness of the high carbon alloy. Work hardening of the surface of low carbon alloys due to the effects of mechanical abrasion were further confirmed by measuring the micro hardness (30 g load) versus distance profiles beneath the wear surface using transverse sections cut and polished through wear scars formed on the wear pin. The sections were also etched to remove any hardening caused by the mechanical polishing process. This data gave some scatter and it was not possible to estimate the depth to which hardening below the wear surface had occurred.

Table 7- *Effect of mechanical polishing and etching on the surface hardness of high and low carbon Co-Cr-Mo alloys*

Alloy	Bulk Hardness, Hv 10 kg As polished	Micro Hardness, mechanically polished, Hv 30g	Micro Hardness, electro etched, Hv 30g
High Carbon	473 ± 5	454 ± 30	493 ± 30
Low carbon	394 ± 2	463 ± 25	375 ± 15

Further evidence of surface hardening due to mechanical abrasion together with the means of estimating the depth of the hardened layer was obtained by measuring the hardness of mechanically polished surfaces at different loads to produce different depths of indentation. Hardness did not vary significantly with load and hence with depth of penetration in the high carbon grade alloy but gave increased levels of hardness in the low carbon grade alloy when the depth of penetration caused by the indentor was less than approximately 12 μm, see Figure 7.

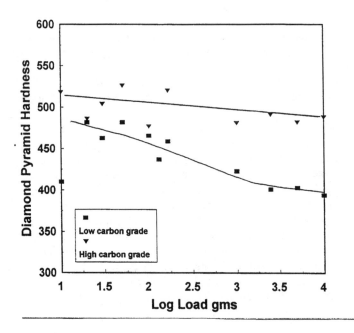

Figure 7 - *Hardness versus load determined by diamond pyramid indentation of a mechanically polished surface in high and low carbon Co-Cr-Mo alloys*

Microstructural Examination

Differences in microstructure between the two types of Co-Cr-Mo alloy also emerged after metallographic examination. The structure of the high carbon grade alloy contained stringers of carbide particles lying along the rolling direction but no such carbides were detected in the low carbon alloy, due to its lower total carbon content, see Figures 2 and 8. Grain size in the high carbon alloy was also much finer than that formed in the low carbon material and there was a notable absence of twins within the grain structure of the high carbon alloy. Twins which essentially equate to stacking faults with the same HCP structure as that of the epsilon phase formed in cobalt alloys at low temperatures were clearly present in the low carbon alloy, see Figure 8.

Figure 8 - *Optical microstructure of low carbon Co-Cr-Mo alloy, mechanically polished and electrolytically etched in 5% aqueous HCl . Normanski Interference Image. Note presence of twins.*

Discussion

The results shown in Table 6 confirm that the high carbon grade alloy gave significantly lower wear rates than the low carbon alloy when loads were sufficiently high to cause contact stresses capable of causing considerable plastic deformation by shear of the surface layers and of producing adhesive wear. This improvement in wear resistance probably stemmed from the increase in hardness produced by the dispersion of fine carbide particles present in the microstructure of the high carbon alloy.

Although there was considerable scatter in the wear test data the results also suggested that very little difference in wear resistance existed between the high and low carbon grade Co-Cr-Mo alloys when tested against a high carbon Co-Cr-Mo alloy disc at low loads and when wear was more determined by abrasion rather than by adhesion. In spite of expectations, the higher hardness of the high carbon grade alloy appears to have produced little benefit to wear resistance at low loads and was in agreement with previous work, [5]. This led to some speculation concerning the possibility that work hardening of the surface in the low carbon alloys may have influenced their wear performance. Shear deformation of the surface under sliding conditions and the resultant plastic deformation was capable of causing work hardening of the surface which appeared to be more pronounced in the low carbon than in the high carbon alloy. The higher M_s temperature, lower stacking fault energy and reduced stability of the FCC phase formed in low carbon Co-Cr-Mo based alloys, [6], possibly compensated for its reduced hardness compared to the high carbon alloy by increasing its capacity to work harden by martensitic transformation on the surface during sliding wear, [7,8]. This effect was supported by the micro hardness test results shown in Table 7 and Figure 7 and was in agreement with previous evidence that work hardening of the surface in cobalt alloys can improve wear properties under low load abrasive wear conditions, [9].

This difference in surface work hardening characteristics and the possible formation of epsilon martensite could also account for the increased tendency for corrosive wear that was seen to occur in the low carbon alloys. Corrosion pitting attack seen on the surface of the low carbon alloys could have occurred by preferential attack of the highly stressed martensitic phase.

Conclusions

1) A process route using vacuum induction melting, electroslag remelting, press and precision forging, and rolling, has been established as a production method for high carbon Co-Cr-Mo with a composition and properties which meet the requirements of ASTM F1537 (warm worked).

2) The microstructure produced by this process route consisted of fine grains of ASTM 10 or finer with fine (approximately 5 μm) primary carbides and discontinuously distributed grain boundary carbides.

3) A limited correlation between strength and nitrogen content was established.

4) The higher carbon grade of wrought ASTM 1537 grade of Co-Cr-Mo alloy showed greater resistance to wear than an equivalent low carbon grade alloy when tested under conditions that create adhesive wear by pin on disc (of high C Co-Cr-Mo) testing at high loads in an *in vitro* corrosive environment. This was attributable to an increase in hardness caused by the presence of carbide particles in the microstructure.

5) Little difference between the wear behaviour of high and low carbon grade alloys was detected when they were tested under low loads against a high carbon disc material, in an *in vitro* environment, and when mainly abrasive wear took place. This was attributed to work hardening of the surface in the low carbon alloy during testing which compensated for any loss in wear resistance that may have arisen from the lower initial hardness displayed by the low carbon alloy.

6) The high carbon grade alloy could possess advantages over the low carbon alloy in relation to its application as a bearing surface for artificial hip and knee replacements. Its greater resistance to adhesive wear at high loads could reduce wear and hence the volume of wear debris produced when high contact stresses tend to exist during the initial bedding in stage experienced by the artificial joint. Reduced tendencies for the high carbon alloy to experience corrosion pitting and corrosive wear in an *in vitro* environment also suggests that the alloy may be more biocompatible and that it may cause less release of metal ions into any surrounding tissues.

Acknowledgements

The support of Firth Rixson Superalloys, Bradford University and California Polytechnic State University is gratefully acknowledged.

References

[1] Lucas L.C., Lemon J.E., Lee J., Dale P., ASTM STP 986 1988 pages 124-136 "Quantitative Characteristics and Performance of Porous Implants".

[2] Jacquet, T., Report on " Commissioning of a Tribometer for the wear testing of high speed Steel Composite Materials", Final Year Undergraduate Project, University of Bradford, 1995.

[3] Hertz, H., *Gesammtlte Werke*, Vol 1, Leipzig, 1895.

[4] Thomas, H. R., Hoersch, V. A., "Stresses due to Pressure of one Elastic Body on Another," Eng. Experimental Station Bulletin 212, Urbana/Champaign, University of Illinois, 1930.

[5] Streicher, R., Semlitsch, M., R Schon, *Proceedings Institute of Mechanical Engineers*. Vol 210, 1996, 223.

[6] Kusoffsky, A., Jansson, Bo., *Calphad*, Vol. 21, No. 3, 1997, 321.

[7] Bhansali, K. J., Miller, A. E., "Wear of Materials", ASME, 1981, 179.

[8] Antony, K. C., Silence, W. L., Proceedings. 5th International. Conference on Erosion by Solid-Liquid Impact, Cambridge University Press, 1979, 6711.

[9] Crook, P., Levy, A. V., ASM Handbook, Vol 18, Friction Lubrication Wear Technology, 1992, 766.

John A. Tesk,[1] Christian E. Johnson,[2] Drago Skrtic,[3] Ming S.Tung,[3] and Stephen Hsu[4]

Amorphous Alloys Containing Cobalt for Orthopaedic Applications

Reference: Tesk, J. A., Johnson, C. E., Skrtic, D., Tung, M., and Hsu, S. **"Amorphous Alloys Containing Cobalt for Orthopaedic Applications,"** *Cobalt-Base Alloys for Biomedical Applications, ASTM STP 1365,* J. A. Disegi, R. L. Kennedy, and R. Pilliar, Eds., American Society for Testing and Materials, West Conshohocken, PA, 1999.

Abstract: Amorphous metal alloys have properties and structures unlike those of their crystalline counterparts. For example, a multiphase crystalline structure may exhibit poor corrosion resistance while the amorphous structure is corrosion resistant. The high hardnesses of some amorphous alloys may make them useful for wear resistant applications. Cobalt-based, electrodeposited alloys may be particularly compatible for producing desirable surfaces on orthopaedic Co-Cr-Mo alloys. Amorphous Co-20P alloy (A-Co-20P) has an as-deposited Knoop hardness number (HK) of *ca* 620. The surface of anodized A-Co-20P has been described elsewhere as predominately phosphorus oxide that may react with water to form an adsorbed hypophosphite layer. Others reported corrosion resistance without pitting. Hence, the potential of A-Co-20P for use as an implant coating to induce bony apposition was evaluated. This evaluation consisted of tests of corrosion resistance and solution chemistry. Another alloy, amorphous Co-Cr-C, was found to have an as-deposited HK of *ca* 690. Heat treatment produced Knoop hardness numbers (HKs) of *ca* 1350, between the HKs of zirconia and alumina. Wear and corrosion resistance are expected to be good, but adherence needs to be assessed.

Keywords: amorphous metals, glassy alloys, orthopaedic joint surfaces, orthopaedic joint surface coatings, corrosion resistant coatings, wear resistant coatings, wear of ultrahigh molecular weight polyethylene, wear resistant implants, bone conductive, bone induction, implant-bone interface

Numerous approaches are currently being employed in attempts to develop orthopaedic biomaterials that offer improved performance over existing materials. For

---

[1] Coordinator, Biomaterials Programs, Polymers Division, National Institute of Standards and Technology (NIST), Gaithersburg, MD 20899.
[2] Physical Scientist, Metallurgy Division, NIST
[3] American Dental Association Health Foundation Paffenbarger Research Center, NIST
[4] Group Leader, Surface Properties, Ceramics Division, NIST

example, numerous approaches for applying calcium-phosphate coatings to orthopaedic implants have been used for the purpose of more rapidly and firmly anchoring them to bone. Also, because wear debris from orthopaedic implants has been implicated as a major cause for the end-of-effective-service of orthopaedic implants [1,2], there is a strong interest in exploring new bearing designs, including new material couples, that may reduce wear debris effects and, hence, help to prolong effective implant service. In this regard, various methods to produce harder, more wear-resistant surfaces on orthopaedic hip implants are being investigated. While modifications of existing materials and mechanical designs must surely be investigated, the potential of new technologies should also be explored for what they may have to offer and for what may be learned toward solving the problem.

This report on work-in-progress is intended to provide results of some tests on the corrosion resistance of amorphous A-Co-20P (atomic fractions are used throughout the paper when specifying particular compositions within an alloy system), and hardness, wear resistance, and adherence of some electrodeposited amorphous Co-Cr-C alloys, Co ≈ (0 to 42) %, Cr ≈ (88 to 14) %, C ≈ (12 to 8) %, respectively {mass fractions, Co ≈ (0 to 50) %, Cr ≈ (97 to 50) %, C ≈ (2 to 3) %}. The Co-Cr-C was electrodeposited onto Co-Cr-Mo alloy meeting the compositional requirements of ASTM Standard F 75 for Co, Cr, & Mo. For the adherence test, a simpler A-Cr-C alloy will be used to obtain an indication of what may be expected from more complex A-Co-Cr-C alloys when they may be used under their own favorable electro-deposited and heat treatment conditions.

Fabrication by Electrodeposition

One way to produce metallic glasses is by electrodeposition, which is a facile method for applying coatings. By this method metallic glasses can be produced that have a variety of structures and properties. They may be produced as alloys of single phase structures or as mixed-phase, layered, structures of which one or all of the layers may be glassy, each with differing, predominate, elemental composition. By control of the electrodeposition times, voltages, voltage waveforms, solutions and other characteristics of processing, the compositions and thicknesses of the layers can be controlled. Brenner et al.[3], Johnson et al.[4], Ratzker et al.[5], and Helfand et al.[6] describe electrochemical processes for fabrication of A-Co-20P alloys; the methods of Ratzker [5] were followed for fabrication of the A-Co-20P alloys evaluated for this paper. The electrodeposition methods reported by Johnson et al.[7,8,] and Soltani [9] were followed for fabrication of the A-Co-Cr-C alloys and A-Cr-C alloys that were tested. There are numerous reports in the patent and other literature that describe the electrodeposition baths and conditions for producing amorphous cobalt-phosphorus alloys, so these will not be reiterated here.

The cobalt-phosphorus and cobalt-chromium-carbon systems are the specific objects of investigation here; carbon is carried into the system from organic components of the electrolyte. Layers in these systems may be homogenized by heat treatments to produce either amorphous materials or materials with noncrystalline structures and/or finely dispersed precipitates. At early stages of treatment, precipitates cannot be detected optically or analyzed by x-ray diffraction; increased hardness indicates their presence.

Composition and Properties

Cobalt-Phosphorus Alloys

In 1993 Ratzker et al.[*5*] described a method for fabricating dental prosthetic frameworks of metallic glasses, composed of preferred composition Co-20P (mass fractions of ca 88 % cobalt and 12 % phosphorus), for use in resin-retained, fixed-partial dentures. The alloy samples had high hardnesses, ranging from Knoop hardness numbers (HKs) of ca 620 to 1100; the latter was obtained following one hour heat treatment in air at 350 °C. The HK tests were made on film cross sections, supported around the edges by 100 µm nickel deposits, at loads of 0.98 N (0.1 kgf). Based on the method of 2[*10*], yield strengths calculated from HKs {(ca 3 MPa)/HK} indicate that these alloys are exceptionally strong, having yield strengths ranging from ca 1,860 MPa to 3,300 MPa (270,000 psi to 478,000 psi). These estimates compare favorably with the minimum yield and tensile strength requirements for cobalt-chromium-molybdenum alloys as given in ASTM specifications F-75 and F-1537, i.e.; 517 MPa and 897 MPa respectively. Such high strength is what was sought for the purpose of fabricating very thin prostheses for conservative tooth preparations. In addition, the corrosion potentials were ca 150 mV versus hydrogen electrode potentials for the as-deposited alloy. In this sense, these are alloys that may be considered noble but their corrosion potentials are lower than those for polished cobalt-chromium-molybdenum dental alloy, which were found to be ca + 700 mV (the dental alloy is virtually the same composition as that required for orthopaedic implant alloys according to ASTM F-75 and ASTM F-1537). The corrosion current of the cobalt-phosphorus alloy was found to comparable to or less that for the dental alloy and there was no evidence of pitting in the deposit.

Ratzker et al.[*5*] described the corrosion current of A-Co-20P as 0.8×10^{-3} cm/year and Co-Cr-Mo as 5.8×10^{-3} cm/year. Ratzker [*11*] also described the dynamic polarization as exhibiting a passive region . Helfand et al.[*6*] also reported on an A-Co-20P alloy that they had prepared by electrodeposition . Helfand et al.[*6*] analyzed the surface of their alloy after it had been subjected to series of anodic polarization treatments, including 10 s at 600 mV, and 1 h at 0 mV (measured against a standard calomel electrode). The latter found evidence for a surface that, after anodic polarization and during immersion in acidic electrolytes, changed to one that was highly enriched in phosphorus, forming a hypophosphite that they speculated inhibits other water molecules from reaching the alloy's surface. Hence, they explain P as inhibiting anodic dissolution by enriching itself on the alloy surface and by the formation of this hypophosphite layer. They termed the anodic polarization as having a region they called "transpassive".

If a stable phosphorus-containing layer of some sort does form on the alloy surface, and if that surface does become truly passive, such a surface may be one to which calcium ions may act as bridges to physiological phosphates in vivo. Under those conditions, it is conceivable that the surface could be osteoconductive and implants could be anchored in bone by use of such amorphous cobalt-phosphorus alloy coating. Therefore, one purpose of investigation was to determine whether the surface layer referred to as passive or "transpassive", formed following anodic polarization of an amorphous Co-20P alloy, would be stable in aqueous solutions.

Cobalt-Phosphorus Alloys, Experimental Procedure

It has been well established that A-Co-20P alloy and the similar Ni-20P [3,6] are enriched in phosphorus following anodic polarization methods treatments. Helfand et al. [6] propose a surface layer that is virtually hypophosphite, $H(H_2PO_2)$. It is further established that the anodic polarization curves display a region that has been interpreted as being passive [11] or "transpassive" [6]. However, implication that this behavior may be passive appears to be questionable in that the anodic current does not fall off as sharply and is not as insensitive to increases in voltage as alloys of the Co-Cr-Mo system that are used for medical and dental purposes. For a binary A-Co-20P alloy to be nontoxic and osteoconductive, the alloy would have to be truly be passive and to have phosphorus incorporated into the surface in a form that is not removed, say by precipitation with calcium ions. Therefore, to investigate whether this capability existed with Co-20P , it was decided to examine the solution behavior of the alloy specimens after several different polarization treatments shown in Table 1.

Specimens were prepared so that portions of the alloy surfaces were masked to prevent surface changes during the polarization treatments. Then tests of the stability of the masked and polarized surfaces were made by monitoring pH changes while the surfaces were in water and in CaP solution (2.1 mmol/L phosphate & 10 mmol/L $CaCl_2$ and at pH = 6.1). The test solutions were also checked afterward for release of cobalt ions. Following this, the specimens were then soaked overnight either in water, 1 mmol/L phosphoric acid or 1 mmol/L sulfuric acid under stirring at a temperature of ≈ 23 °C, to determine whether all cobalt that may exist in the surface could be removed (solutions were in excess volume such that there was ≈ 1 L solution for each 2 cm^2 of Co-20P surface). After the overnight soakings, the specimens were washed with distilled water and they were once again tested in water and in calcium phosphate solution (previously described) to determine whether additional cobalt would be released. In this procedure, one drop of solution, either water or calcium phosphate solution, was put on the surface and a combination pH electrode was used to continuously follow the pH of

Table 1
Polarization Conditions

SOLUTION	CURRENT DENSITY	TIME
0.2/3 mol/L H_3PO_4,	2.5 mA/cm^2	1 min
0.2/3 mol/L H_3PO_4,	6.5 mA/cm^2	2 min
0.2 mol/L HCl.,	6.5 mA/cm^2	2 min

the solution. Afterward, a test for the presence of cobalt ions in the solution was made using cobalt test paper (limit of sensitivity: Co, 25 mg/L; Gallard-Schlesinger Chemical Mfg. Corp. [12]). This method can characterize the solution tendency of the surface. Changes in pH follow the dissolution behavior in water, and the dissolution and precipitation behavior in the calcium phosphate solution. Dissolution of basic cobalt compounds increases the pH of the solutions, conversely, the precipitation of basic apatite from the calcium phosphate solution (metastable with respect to hydroxyapatite) in the absence of release of basic cobalt will decrease the pH of the solution [*13*].

Cobalt-Phosphorus Alloys, Experimental Results

Figure 1 shows typical increases in pH as function of time in the calcium phosphate solution after polarization (even faster increases in pH of water were observed; typically pH increased from 6.4 to 7.6 in 3 h). These increases in pH indicate the dissolution of the basic cobalt compound into the water and the calcium phosphate solution. While the precipitation may occur in the calcium phosphate solution, it was not observed by the pH measurement because of the dominance of the dissolution process. Within 50 min, cobalt ions were detected both in the water and calcium phosphate solution using the cobalt test paper. Further, after soaking in water, phosphoric acid, or sulfuric acid over night, with stirring, the release of cobalt ions was still observed.

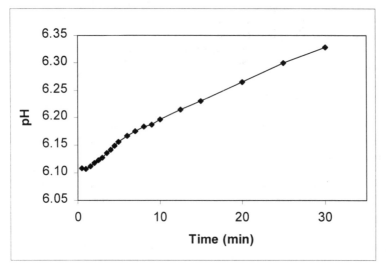

Fig. 1 *Typical increase in pH as a function of time in the calcium phosphate solution. The pH is for a drop of solution placed on a previously polarized surface of a Co-20P alloy specimen.*

Cobalt-Phosphorus Alloys, Discussion and Conclusions

The continuous dissolution of cobalt from previously anodically polarized amorphous Co-20P alloy, with no apparent cessation of that behavior, is proof that surfaces of that alloy are not passive (this occurred in acids, water, or acidified calcium phosphate solutions). The evidence presented in the literature of a phosphorus enriched surface cannot be discounted, but because the surface does not passivate, amorphous Co-20P is not suitable for use as an implant material. However, there are reports in the literature that amorphous Ni-Cr-P alloys can be passivated and that the presence of phosphorus in the surface layer contributes to this passive behavior. By analogy with the nickel alloys, it is reasonable to expect that similar behavior could be obtained if an amorphous Co-Cr-P alloy could be obtained by adding chromium to the system to make it more corrosion resistant, with both Cr and P contributing to a corrosion resistant surface. If this analogy carries forth, then there remains the possibility of fabricating implantable amorphous Co-Cr-P alloys that would also have the property of osseous integration by direct bonding to bone, through calcium linkages to phosphorus in the passive layer on the surface of the alloy.

Cobalt-Chromium-Carbon Alloys

Because of the high hardnesses of as-deposited amorphous chromium-carbon (HK ≈ 800) [8] the electrodeposited amorphous cobalt-chromium-carbon alloys were chosen for investigation as their hardnesses were expected to be similarly high. The potential advantages of the latter alloys for use as coatings for orthopaedic bearing surfaces and instruments lie in: 1) heat treated condition hardnesses that, based on the hardnesses attainable from heat treated amorphous chromium-carbon, are expected to lie between those of high purity zirconia and high purity alumina and which should, therefore, provide highly wear resistant surfaces, 2) chemical compositions that should impart good corrosion resistance, 3) electrochemical deposition processing that should be a facile method for producing uniform coatings on simple geometry bearing surfaces. Therefore, the objective of this section of this manuscript was to investigate: 1) whether, in fact, high hardnesses, similar to those attainable from heat treated amorphous chromium-carbon, could be attained, 2) and whether the electrodeposited coatings had adherence and wear resistance sufficient to render them potentially useful for service as wear-resistant coatings for orthopaedic implant surfaces and instruments.

Cobalt-Chromium-Carbon Alloys, Experimental Procedure

In essence, the alloys were deposited from electrolytes containing di-valent cobalt and tri-valent chromium ions, with voltages that alternated back and forth between that needed for depositing cobalt to that needed for depositing chromium. This results in a structure composed of amorphous Co-rich and Cr-rich layers (Fig. 2). During such a procedure, the length of time spent at each voltage determines the thickness of a layer. As deposition of each layer proceeds, carbon is known to be carried into the chromium deposits from the organic constituents of the electrolyte [3]. Details of the electrodeposition procedures have been previously described elsewhere [7,8,9]. X-ray diffraction using Cu K_α radiation was employed to assess the amorphous nature of the deposits.

The thicknesses of the layers deposited for fabrication of specimens for hardness measurements were *ca* 1 μm for the chromium-based layer and *ca* 0.4 μm for the cobalt-based layer. The HKs were taken from film cross sections, supported around the edges by 100 μm thick nickel deposits, at loads of 0.245 N (.025 kgf) , Fig. 3. Prior to indentation the surfaces of the cross sections were polished with 0.05 μm alumina in distilled water. After hardnesses of as-deposited specimens were obtained, specimens were heat treated

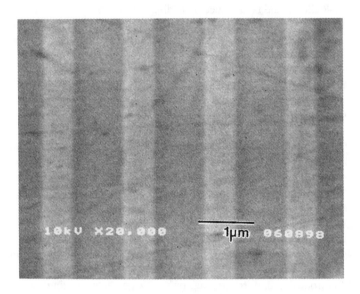

Fig. 2 *Layered structure of Co-Cr-C alloy with layers, 0.9 μm for Cr, 0.6 μm for Co, .*
Layers as thin as 10 nm or less can be produced (Magnification = 20000 X).

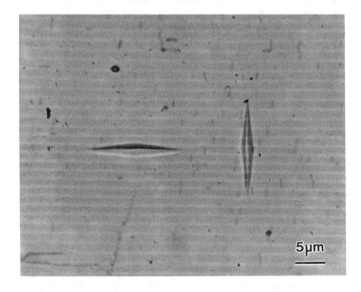

Fig. 3 *As-deposited layered structure of cobalt-chromium-carbon with Knoop hardness*
indentations (Magnification = 2000 X)

for 1 h at 600 °C in a vacuum at a pressure of 13.3 μPa (1 x 10^{-7} Torr). This treatment is known to produce the maximum hardness of ca 1850 from the transformation of the amorphous Cr-C alloys to a crystalline structure that results in clearly visible, dispersed, carbide precipitates in the chromium (in those alloys, treatment at 500 °C for 1 h results in nearly the same hardness, ca 1750 HKs, but precipitation cannot be resolved by either optical nor x-ray diffraction methods). The heat treatment of Co-Cr-C at 600 °C was similarly expected to reveal carbides in the chromium while precipitation in the cobalt would be taken as evidence of carbon also having been present in the amorphous cobalt layer. Following the 600 °C treatment, hardness indentations were taken both parallel to and perpendicular to the layer directions (attempts were made to keep the major axes of layer-parallel indents within a single layer and without termination at interfaces but this could not always be done). After the hardness tests, the specimens were etched for 2 min with a solution of 2.67 mol/L $K_3Fe(CN_6)$ and 0.4 mol/L KOH in distilled water, rinsed with distilled water, and dried. They were then observed optically at a magnification of 2000 X to determine whether precipitation had occurred, how uniform precipitation was within the chromium and cobalt layers, and whether there was evidence of diffusion between layers.

Specimen preparations for the wear and adherence tests have not yet been completed as there was an extensive shut-down of the electrodeposition laboratory that

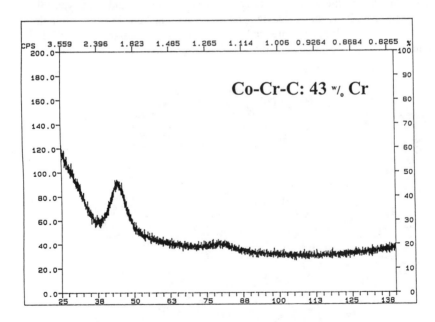

Fig. 4 *X-ray diffraction showing the broad pattern from glassy cobalt-chromium-carbon multilayer deposit, mass fraction Cr = 43 %. Counting Rate (s^{-1}), standard uncertainty = 2 s^{-1}, vs the diffraction angle, 2θ, (resolution 0.01°).*

has inhibited this intended research. At the time of this submission of this manuscript, the laboratory is in the final stages of renovation, after which specimens will be fabricated for these tests.

Cobalt-Chromium-Carbon Alloys, Experimental Results

The X-ray diffraction patterns of as-deposited alloy were characteristic of an amorphous structure (Fig. 4). Distilled water was needed for the polishing procedures used to prepare surfaces for microscopy and hardness indentations as tap water had an etching effect on the cobalt layer, both before and after the 600 °C heat treatment.

The amorphous Co-Cr-C has an as-deposited HK of (693 ± 7.6), with a value of (692.4 ± 5.4) from measurements along the layer directions, and (694.8 ± 14.4) across the layers ($\pm$ is the estimated standard uncertainty of the measurements). Subsequent heat treatment at 600 °C resulted in HKs of *ca* (1300 to 1400). With the indenter tip in the chromium layer, the HK was (1300 ± 24) when the long axis of the indenter was parallel to the layer; when the long axis of the indenter was perpendicular to the layer the HK was (1407 ± 21). In no instance was delamination between layers found to have been caused by the HK indentations.

After the 600 °C treatment, cracks perpendicular to the layers could be observed in the chromium layers; these cracks were generally blunted by and did not traverse the adjacent cobalt layer. Subsequent etching revealed carbides had precipitated within the chromium layers and carbides appeared to had begun forming within the cobalt. The boundaries between the layers became more diffuse (Fig. 5) than they had been for the

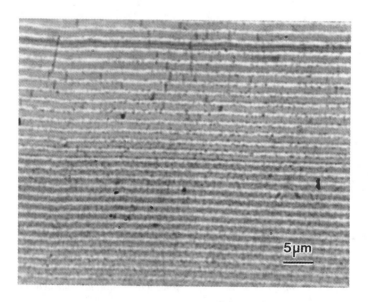

Fig. 5   *Co-Cr-C multilayer showing carbides after 600 °C heat treatment.*

as-deposited structure, indicating that diffusion was also taking place between layers (for an example of layer boundary structure prior to heat treatment see Fig. 2). The extent of the diffuse boundary could be estimated from the Fig. 5 photomicrograph as *ca* 0.2 µm.

Cobalt-Chromium-Carbon Alloys, Discussion and Conclusions

The resistance to delamination under indentation is important because it provides evidence that the layers may be bonded sufficiently to resist delamination during service if the alloys were used for implant coatings. Also, the high HKs appear to be the result of the development of very fine precipitates from transformations from amorphous deposits. Prior experience with amorphous chromium carbon indicates that high HKs may be achieved before carbides are either visible by optical microscopy at magnifications of 2000 X or detectable by X-ray diffraction. Such deposits are expected to have only small asperities, this could be important for wear couples for which asperities are a source of decreased wear resistance .

Finally, because elements can often be co-electrodeposited as alloys by choosing electrochemical potentials between those of either element, this approach will also be investigated. An amorphous Co-Cr-C alloy deposit may be advantageous for helping to ensure the attainment of corrosion-resistant, heat-treatable, adherent coatings.

While the data have been limited to structure and hardness determinations thus far there are a number of conclusions that may be drawn from the results. These are:

1) X-ray diffraction reveals the as-deposited layers of Cr and Co as both being amorphous.

2) Heat treated deposits develop HKs that exceed those of zirconia and significantly approach those of alumina. These high HKs indicate potential usefulness for increasing the wear resistance of orthopaedic joints and instruments.

3) The cobalt layer is not corrosion resistant, indicating that chromium is not present in sufficient quantity in the as-deposited Co layer and that the layered deposits used are too thick to have allowed sufficient diffusion of chromium into the cobalt to impart corrosion resistance.

4) The cobalt and chromium layers appear to interdiffuse at their interfaces to the extent of ca 0.2 µm.

5) Other experience in this laboratory is that electrodeposition can deposit alternating layers of Co-C and Cr-C, with layer thicknesses controllable from ≈ 2 µm to ≈ 10 nm (≈ 40 atoms). Therefore, it is likely that thin, layered structures of cobalt-chromium-carbon can interdiffuse sufficiently to provide good corrosion resistance throughout heat treated deposits.

6) Layers are bonded together well enough to resist delamination under indentation.

7) Adherence and wear tests are needed to confirm whether or not useful properties can be attained with deposits on orthopaedic bearing alloys.

Acknowledgment
The authors gratefully acknowledge the assistance of Jasper P. Mullen in generation of X-

ray diffraction and compositional analysis.

References

[1]   Peters Jr., P.C., Engh, G.A., Dwyer, K.A., and Vinh, T.N., "Osteolysis After Total Knee Arthoplasty Without Cement," *Journal of Bone and Joint Surgery*, July, 1992, Vol. 74 No. 6, pp. 864-886.

[2]   Schmalzreid, T. P., Jasty, M., and Harris, W.H., Preiprosthetic Bone Loss in Total Hip Arthroplasty: "The Role of Polyethylene Wear Debris and the Concept of the Effective Joint Space," *Journal of Bone and Joint Surgery* , July, 1992, Vol. 74A No. 6, pp. 849- 863.

[3]   Brenner, A., Couch, D.E., and Williams, E.K., "Electrodeposition of Alloys of Phosphorus with Nickel or Cobalt," *Journal of Research, National Bureau of Standards*, Janurary, 1950, Vol. 44, pp. 109-122.

[4]   Johnson, C.E., Mullen, J.L., and Lashmore, D.S.,"Corrosion Tests of Electrodeposited Coatings Resistant to Boiling Phosphoric Acid," Proceedings of 37th Meeting of the Mechanical Failure Prevention Group, Cambridge University Press, 1884,

[5]   Ratzker, M., Lashmore, D.S., and Tesk, J.A., U.S. Patent No. 5,316,650, Electroforming of Metallic Glasses for Dental Applications,   31 May, 1994.

[6]   Helfand,M.A., Clayton, C.R., Ciegle, R.B., and Sorenson, N.R., "The Role of P in Anodic Inhibition of an Amorphous Co-20P Alloy in Acidic Electrolytes," *Journal of the Electrochemical Society*, August, 1992, Vol. 139 No.8, pp. 2121-2127.

[7]   Johnson, C.E., Lashmore, D.S., and Soltani, E., U.S. Patent No. 5,415,763, Methods and Electrolyte Compositions for Electrodepositing Chromium Coatings, 16 May, 1995.

[8]   Johnson, C.E., Lashmore, D.S., and Soltani, E., U.S. Patent 5,672,262, Methods and Electrolyte Compositions for Electrodepositing Metal-Carbon Alloys, 30 Sept., 1997.

[9]   Soltani, E.C., Cobalt-Chromium Multilayer Alloys, Thesis submitted for degree of Master of Science, The University of Maryland, 1992.

[10]   Tabor, D., The Hardness and Strength of Metals, *Journal of the Institute of Metals*, Vol. 79: pp. 1-18, 1951.

[11]   Ratzker, M., unpublished research, NIST, 1993.

[12]   Commercial material identified only for specifying experimental procedure, not endorsed by NIST nor claimed to be the best for the intended purpose.

[13]   Tung, M.S., Bowen, H.J., Derkson, G.D., and Pashley, D.H., "The Effects of Calcium Phosphate Solutions on Dentin Permeability," *Journal of Endodontics*, August, 1993, Vol.19 No 8, pp. 383-387.

Mechanical Properties

Barbara S. Becker,[1] and John D. Bolton[1]

Effect of Powder Morphology and Sintering Atmosphere on the Structure-Property Relationships in PM Processed Co-Cr-Mo Alloys Intended for Surgical Implants

Reference: Becker, B.S., Bolton, J.D., **"Effect of Powder Morphology and Sintering Atmosphere on the Structure-Property Relationships in PM Processed Co-Cr-Mo Alloys Intended for Surgical Implants,"** *Cobalt-Base Alloys for Biomedical Applications, ASTM STP 1365,* J. A. Disegi, R. L. Kennedy, R. Pilliar, Eds., American Society for Testing and Materials, West Conshohocken, PA, 1999.

Abstract: This study analysed the possibility of producing biocompatible, porous functionally graded metallic acetabular cups using powder metallurgy techniques, where a porous surface provided a substrate for incorporating a polymeric compliant layer which was supported by a rigid inner core. Two different Co-Cr-Mo powders were investigated, a water atomised powder and a novel bimodal powder; each had a different powder morphology leading to differences in compressibility and sintering behaviour. Suitable combinations of compaction pressure and sintering temperature were used to produce porous materials containing 5 - 33% total porosity. Moreover, sintering was conducted in three different atmospheres; under vacuum, in flowing argon and in a flowing molecular mixture of 75%H_2/ 25% N_2. The resulting alloys were subjected to mechanical tests including hardness, tensile strength and ductility. Variations in pore morphology led to differences in surface chromium, bulk nitrogen and carbon content which resulted in marked differences in properties, although the lower the bulk porosity the better the properties exhibited.

Keywords: Powder Metallurgy, Co-Cr-Mo Alloys, Sintering Behaviour, Corrosion Behaviour, Mechanical Properties

Currently, the accepted average lifetime of an artificial hip joint is ten years which, in the main, has been attributed to joint loosening due to foreign body tissue reactions to polymeric wear debris from the ultra high molecular weight polyethylene acetabular cup, against which the metal or ceramic femoral head articulates.

---

[1] Research Fellow and Reader in Medical Engineering, respectively, Engineering Materials Research Unit, Dept. of Mechanical and Medical Engineering, University of Bradford, Richmond Road, Bradford, West Yorkshire, BD7 1DP, UK.

Recent research [*1*] has shown that incorporating a polymeric compliant layer in to the bearing surface of a fully dense metallic acetabular cup can give rise to a fluid film between the cup and the femoral head through elastohydrodynamic lubrication. This fluid film separates the bearing surfaces, effectively eliminating wear during normal walking. Finding a suitable metal substrate/compliant layer combination could extend the acceptable life of an artificial hip joint to 25 years or more, permitting these devices to be used in younger, more active patients. Prior to this work, polymeric compliant layers had been incorporated into a fully dense cup substrate using four fixing locations but this cup only lasted a few weeks testing on a hip-joint simulator. Powder metallurgy techniques offer the possibility of creating a near-net part that has hundreds of surface pores providing suitable location sites in to which the polymeric compliant layer could be bonded. By combining a porous surface with a dense core (through developing a functionally graded component), the cup would have sufficient mechanical properties to withstand the physiological loads.

Commercially available Co-Cr-Mo powders, in a gas atomised or Plasma Rotated Electrode Process (PREP) form, are currently used to produce fully dense surgical implants [*2,3,4*] but neither the powders themselves nor the consolidation processes can be easily altered to produce porous parts. Water atomised Co-Cr-Mo powders could be used but are not commercially available, despite a number of research projects being undertaken using these powders [*5, 6*]. Workers claimed that the water atomised powders had low oxygen contents and could be compacted at reasonable pressures with suitable green densities, but very little data exists in the literature. In 1968, Hirschhorn and Reynolds [*5*] successfully manufactured porous femoral stems from water atomised cobalt based alloys. They developed a femoral stem with varying porosity from a dense metal core to an outer open pore network by isostatically pressing the powder using a wet bag technique. Their work showed that boney in-growth into the porous stem occurred while the graded structure prevented stress shielding. Although Hirschhorn argued that his powder manufactured femoral stems had adequate mechanical properties and corrosion resistance suitable for this application, others suggested otherwise [*7*] and especially doubted the fatigue properties.

Interestingly, Hirschhorn and Reynolds speculated that a porous material could be impregnated with lubricating fluids, or plastics, to reduce the wear rates in applications where a porous material might be employed as an articulating surface. Some years later Pilliar [*8*] reported on a porous coated femoral head component in to which a biologically-compatible hydrophilic polyurethane was moulded to form a more energy-absorbing hip replacement. No further development work or clinical trials for either of these two concepts was found in the literature.

Recent research takes Hirschhorn's porous stem concept a stage further, by developing functionally graded materials, manufactured either by gas pressure combustion sintering [*9*] or by plasma spraying techniques, to lay down successively decreasing particle size fractions onto a wrought substrate [*10*]. The resulting graded porosity materials have found application as dental tooth roots [*11*]. Functionally graded type materials, produced using conventional powder metallurgy techniques, offer the ideal solution to the development of an acetabular cup containing a compliant layer. The level of porosity can be graded: from a highly porous surface layer, suitable for polymer impregnation, to a dense core, giving the component suitable strength to withstand the physiological loadings.

The aim of this work was to establish the feasibility of developing a metallic porosity graded acetabular cup into which a polymeric compliant layer could be incorporated to replace the UHMWPE cup currently available. The approach was to produce porous materials containing a single porosity, using Co-Cr-Mo alloy powders by conventional cold compaction and sintering techniques. A study was made of the effect of sintering conditions and the inherent porosity on the microstructures of the alloys, together with the effect of porosity on mechanical properties such as hardness, tensile strength and ductility.

Experimental Method

Materials

Water atomised Co-Cr-Mo powder was produced by INETI, Lisbon, Portugal and Scientific Metal Powders Ltd., Sheffield, South Yorkshire, UK using ASTM Standard Specification for Cast Cobalt-Chromium-Molybdenum Alloy for Surgical Implant Applications (F75 - 87) bar stock (with modified carbon levels), supplied by Howmedica International, Limerick, Ireland. Both batches of water atomised powder contained a high proportion of large spherical shaped particles. After sieving, three powder size fractions were used in subsequent sintering trials and mechanical tests: a coarse powder fraction (353-106μm), a medium powder fraction (75-45μm), and a fine particle fraction (<38μm). Each fraction was vacuum annealed at 900°C for two hours to soften the powder and to reduce any oxides present. To create the novel bimodal powder, Co-Cr-Mo Plasma Rotated Electrode Process (PREP) powder, supplied by Nuclear Metals Inc., was ball milled for 64 hours in acetone using WC balls. After drying in an oven and vacuum annealing at 900°C, 50 weight percent of the ball milled powder was dry blended with 50 weight percent of coarse fraction (353-106μm) water atomised powder. (Table 1) indicates the composition of the ball milled and the all particle water atomised powder after annealing.

Table 1 – *Bulk chemical compositions in weight percents (wt%) of the water atomised powder and the ball milled PREP powder after annealing*

Powder Type	C	N	O	Cr	Mo	Ni	Si	Mn	Fe	Co
Ball Milled - After Annealing	0.41	0.075	0.655	26.24	4.63	3.45	1.18	0.63	6.24	Bal
Water Atomised - After Annealing	0.32	0.15	0.470	27.50	5.80	0.60	0.00	0.63	0.76	Bal

Note the high nickel and iron contents of the ball milled PREP powder picked up from the stainless steel jars.

Compaction and Sintering

Both powders were cold compacted at pressures between 470-1235 MPa, using a manual uniaxial press with carbide insert dies. Neither integral lubricant nor die-wall lubricant was used. Sintering of compacted samples was performed in either one of three different sintering atmospheres: vacuum, flowing argon or a flowing molecular mixture of 75%H_2/25%N_2, (both gas atmospheres were at a dewpoint of less than -55°C). Sintering temperatures were varied between 1150 and 1350°C for a single time period of 60 minutes. All compacts were heated at 10°C min^{-1} to the sintering temperature. Gas atmosphere sintered samples were cooled naturally in the furnace from the holding temperature. Vacuum sintered samples were cooled at 5°C min^{-1} from their sintering temperature until they reached 400°C and were then allowed to cool naturally to ambient. Green densities were determined by weighing and by calculating volume from measured specimen dimensions. Sintered densities and porosity characteristics were determined using a xylene impregnation technique which was a modification of ASTM Test Method for Density and Interconnected Porosity of Sintered Powder Metal Structural Parts and Oil-Impregnated Bearings (B328-73) adopted from Raghu *et al* [12].

After the compaction and sintering trials, the porosity of each powder type generally ranged between 5-35% total porosity with interconnected porosity of 0.9-26%. (Table 2) lists the processing details to create three different porosity levels for both the water atomised powder and the bimodal powder. Once the compaction pressure and sintering temperature were set, each powder porosity fraction was sintered in the three different atmospheres to generate samples for mechanical testing.

Samples for metallographic examination were mounted in low viscosity epoxy resins under vacuum and polished by grinding to 1200 SiC grit, through 6μm and 1μm diamond paste, finally lapping with a 0.05μm silica colloidal suspension. Cross sections were then examined using both optical and Scanning Electron Microscopy (SEM) methods, including Energy Dispersive Analysis (EDAX). Note that throughout this work ASTM Standard Specification for Cast Cobalt-Chromium-Molybdenum Alloy for Surgical Implant Applications (F75 - 87) bar stock (with modified carbon levels), supplied by Howmedica International, Limerick, Ireland, was used for comparison.

Table 2 – *Material processing parameters*
for bimodal (BM) and water atomised (WA) powders

Particle Fraction	Nominal Total Porosity (%)	Compaction Pressure (MPa)	Sintering Temperature (°C)
BMC	33	470	1150
BMM	25	617	1250
BMF	10	1080	1350
WAC	33	770	1150
WAM	20	770	1350
WAF	5	1235	1350

Note: C = coarse total porosity (> 30%), M = medium total porosity (~20%), F = fine total porosity (<10%)

Mechanical Testing

Samples for Vickers Hardness measurements were mounted in bakelite (as impregnation using epoxy resins supported the pores during testing) and polished to 1200 SiC grit. Testing was undertaken using a 10kg load. Micro-hardness values were determined using a Reichert Micro-hardness Tester with a loading of 50g.

Ultimate tensile strengths (UTS), 0.2% yield strengths and ductility were determined from tensile testing where 56 x 14 x 10 mm blanks were machined to a gauge diameter of 6 mm over a length of 25 mm and tested on a computer controlled Instron 5568. Initially a 12.5mm extensometer was used up to 0.1 % strain permitting 0.2% yield strengths to be determined. The extensometer was then removed and the sample tested to fracture, enabling UTS and % elongation to be determined.

Results and Discussion

Powder Characterisation

The water atomised powder consisted of roughly spherical particles, (Figure 1a). Sphericity was linked to powder size, the smaller particles appearing more rounded than those particles greater than 106μm in size.

Ball milling the PREP powder transformed this highly spherical powder into small flaky particles, the majority of which were < 38μm in size, (Figure 1b). By combining the large irregularly shaped particles of the water atomised powder with the fine flaky powder of the ball milled PREP powder, a truly bimodal powder was created.

Some of the larger water atomised particles were hollow which are traditionally removed using special aerodynamic techniques such as cyclone separation. However, when combined with the ball milled powder these did not appear to pose a great problem; firstly the sintered structures showed that the fine flaky powder had ingressed into these hollows during compaction, while some of the hollow particles near the outer surfaces of the compact collapsed. Secondly, pore closure was aided by the formation of a liquid phase especially when bimodal powders were sintered at temperatures above 1300°C.

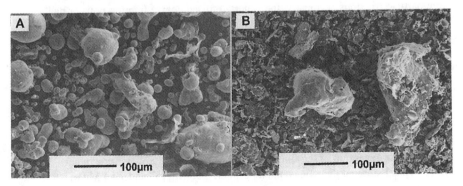

Figure 1 – *SEM Micrographs of powders: (A) Water Atomised, (B) Bimodal*

Both water atomised and ball milled powders proved extremely difficult to press resulting in low green densities at comparatively high compaction pressures. For the ball milled powders this was attributed to poor flow properties leading to poor die packing while the largely spherical nature of the water atomised powder accounted for the poor compaction characteristics exhibited by this powder. In particular the fine fraction water atomised powder could not be compacted below 617MPa (Figure 2) as it crumbled on ejection from the die.

The bimodal powder compacted well over a range of pressures although samples compacted below 470MPa were very fragile and therefore deemed unsuitable for further work. A small over-pressure was exerted on the punches to prevent delamination of the compact during ejection from the die, caused by the release of residual stresses generated in the powder particles during compaction.

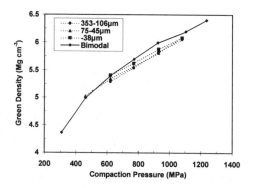

Figure 2 – *Compaction characteristics of water atomised and bimodal powders*

Sintering Characteristics

Effect of Powder Particle Size and Compaction Pressure – For the water atomised powders only minor differences appeared in the sintering behaviour observed between the different particle fractions. Typically, the water atomised coarse fraction powder had a high porosity at low compaction pressures, (Figure 3a) whilst the fine fraction water atomised powder had a low porosity when compacted at high pressures (Figure 3b). As the compaction pressure increased, the higher green densities led to higher sintered densities.

The minimum and maximum sintered densities for the coarse powder were 5.59 Mg m^{-3} and 6.41 Mg m^{-3}, for the medium powder were 5.76 Mg m^{-3} to 6.71 Mg m^{-3} and for the fine powder were 6.18 Mg m^{-3} and 7.31 Mg m^{-3}.

In contrast, higher sintered densities were obtained for the bimodal powder when pressed at lower compaction pressures than those required for the water atomised powders with the exception of the fine fraction water atomised powder pressed at 770MPa. Typical sintered densities ranged between 5.55 Mg m^{-3} and 7.54 Mg m^{-3}.

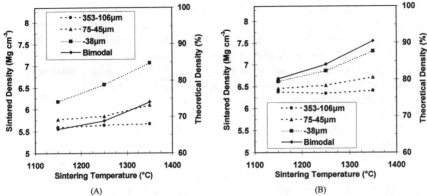

Figure 3 – *Sintering characteristics under vacuum: (A) water atomised at 770MPa, (note bimodal at 470MPa), (B) water atomised at 1235MPa, (note bimodal at 1080MPa)*

Effect of Sintering Atmosphere on Porosity – Water atomised Co-Cr-Mo powders showed considerable sensitivity to sintering atmosphere especially for the fine powder fraction sintered in argon. With this powder the sintered density varied by approximately 10% between the three different atmospheres (Figure 4a). Neither sintering under vacuum nor sintering in a flowing molecular mixture of 75%H_2/25% N_2 significantly altered the final sintered densities (and hence total porosity levels) for this powder. The more open pore geometry may have enhanced the movement of the atmosphere through the sample and assisted in reducing the surface oxides on the powder particles.

Nominal porosity values of the bimodal powders showed little sensitivity to sintering atmosphere until compacting at higher pressures (> 1080MPa) and sintering at temperatures above 1250°C, (Figure 4b).

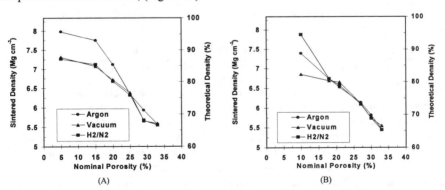

Figure 4 – *Effect of Sintering Atmosphere on nominal porosity: (A) Water Atomised Powders, (B) Bimodal Powders*

Now there was a marked difference in the final sintered densities which indicated that sintering under vacuum produced the lowest sintered densities whilst sintering in a flowing molecular mixture of 75%H_2/25% N_2 exhibited the highest sintered densities. In this case, the sintered density varied by 15 % between the three atmospheres.

In general, sintering atmosphere affected both powders in a similar manner; at lower temperatures (high porosity) sintering was by solid state diffusion leading to only minor variations in the nominal porosity. At higher temperatures (in excess of 1300°C) the sintering atmosphere appeared to promote or inhibit some form of liquid phase sintering; both argon and nitrogen containing atmospheres appeared to lower the solidus temperature allowing a greater amount of liquid phase to form.

Effect of Sintering Atmosphere on Microstructure – After sintering in the three different atmospheres, both powders showed significant bulk compositional variations, (Table 3). The oxygen content was reduced in all cases (compared to Table 1) as was iron in the water atomised samples and iron, nickel and silicon in the bimodal powder samples. Sintering under vacuum resulted in a significant loss of carbon due to the formation of carbon monoxide during sintering leading to surface decarburisation. Whilst sintering in a flowing molecular mixture of 75%H_2/25%N_2 produced an increase in nitrogen content in both powders: 0.7 wt % in the bimodal powders and 1.1 wt % in the water atomised powders. The higher nitrogen content in the water atomised powders samples was probably linked to the more open pore geometry of these samples.

Table 3 – *Bulk chemical compositions in weight percents (wt%) of all particle powder water atomised (WA) and bimodal (BM) samples sintered at 1150°C*

Powder Type	C	N	O	Cr	Mo	Ni	Si	Mn	Fe	Co
WA - AP (Argon)	0.31	0.016	0.367	27.47	5.73	0.54	-	0.59	0.28	Bal
WA - AP (Vacuum)	0.011	0.003	0.148	26.02	5.71	0.50	-	0.53	0.24	Bal
WA - AP (H_2/N_2)	0.013	1.13	0.069	27.38	5.74	0.52	-	0.60	0.20	Bal
BM - (Argon)	0.142	0.008	0.188	26.31	5.37	1.84	0.24	0.62	3.34	Bal
BM - (Vacuum)	0.028	0.005	0.07	25.54	5.32	1.93	0.48	0.51	3.24	Bal
BM - (H_2/N_2)	0.036	0.68	0.078	24.87	5.15	1.88	0.26	0.58	3.24	Bal

Note: Water atomised powder compacted at 770MPa and bimodal powder compacted at 470MPa

Microstructural differences between the water atomised and bimodal powders, sintered in the three different atmospheres, was very marked. Under vacuum, small carbides were seen sited at the grain boundaries, the amount and size of these carbides decreased as the sintering temperature increased. Generally more carbides were seen in the water atomised samples compared to the bimodal samples. EDAX analysis of the surface chromium levels of water atomised (all particle powder) revealed that significant chromium loss occurred up to the first 20μm of the compact due to the dissociation of oxides, such as Cr_2O_3, present on the outer surface of the powder particles. As the sintering temperature increased, the amount and depth of the chromium loss increased, accounting for a 40% increase in loss at a depth of 90μm when the temperature was increased from 1150°C to 1350°C. Therefore, carbides normally associated with Co-Cr-Mo alloys, e.g. M_7C_3 and

$M_{23}C_6$ were less apparent than expected because of chromium volatilisation and decarburisation during sintering.

Argon sintering also revealed significant differences in the microstructure of the two powders. The coarse porosity samples exhibited a small grain size with a high volumetric quantity of discrete carbides sited around grain boundaries. The number of carbides was greater and the grain size was smaller than equivalent samples sintered under vacuum. EDAX analysis confirmed that negligible chromium volatilisation or decarburisation had taken place so that extensive carbide formation was possible. Thus the argon atmosphere protected the free surfaces of the samples. At higher temperatures these carbides appeared to have redissolved in solution since none were visible in the medium porosity samples. With the fine porosity samples the most significant microstructural difference was seen. The bimodal samples exhibited large equiaxed grains surrounded by carbide eutectics. Although large equiaxed grains were also present in the fine porosity water atomised samples, the carbide eutectics were only found in a few isolated areas. Larger volumetric quantities of carbide were evident in the bimodal samples compared to the water atomised samples despite being sintered at the same temperature which was possibly associated with smaller powder particle size, since both chromium and carbon levels were higher for the water atomised powder.

Previous research on the eutectic structures formed at the grain boundaries in the as-cast porous coated femoral stem had identified a brittle $M_{23}C_6$ and Co eutectic which had a solidus temperature of $\approx 1235°C$ [13]. However, EDAX analysis of the carbide eutectics found in the fine porosity water atomised and fine porosity bimodal samples showed that two types of carbide were present, one a chromium rich carbide associated with a high oxygen content and one a molybdenum-silicon rich carbide which could have possibly been sigma phase. More detailed sintering studies suggested that these eutectic structures began to form at 1335°C for the bimodal samples and at 1345°C for the water atomised samples, [14]. Sintering both powders in argon resulted in local incipient fusion around the carbides giving rise to a transient eutectic liquid consisting of γ + either M_7C_3 or $M_{23}C_6$.

Sintering both powders in a flowing molecular mixture of $75\%H_2/25\%N_2$ significantly altered the microstructures again. The coarsest porosity bimodal samples had small discrete chromium nitrides at grain boundaries (Figure 5d), confirmed by EDAX analysis, whilst in the coarse porosity water atomised samples large cellular/lamellar chromium nitride precipitates were clearly visible, (Figure 5a). These latter precipitates had formed preferentially along prior particle boundaries, where the surface was in contact with the atmosphere, although at higher sintering temperatures these cellular nitrides migrated from prior particle boundaries to grain boundaries, (Figures 5b and 5c). Discontinuous precipitation of chromium nitrides has been noted in cast Co-29Cr-6Mo alloys deliberately alloyed with nitrogen. The nature of the nitrides was found by X-ray diffraction to be of the form β-Cr_2N and Cr_2 (C, N), [15].

As the amount of porosity decreased so the nature of the precipitate changed; for the bimodal samples, discrete chromium nitride precipitates which had initially formed at the grain boundaries, grew to a cellular/lamellar structure at sintering temperatures above 1250°C (Figure 5e), disappearing completely when the sintering temperature was increased above 1300°C, (Figure 5f).

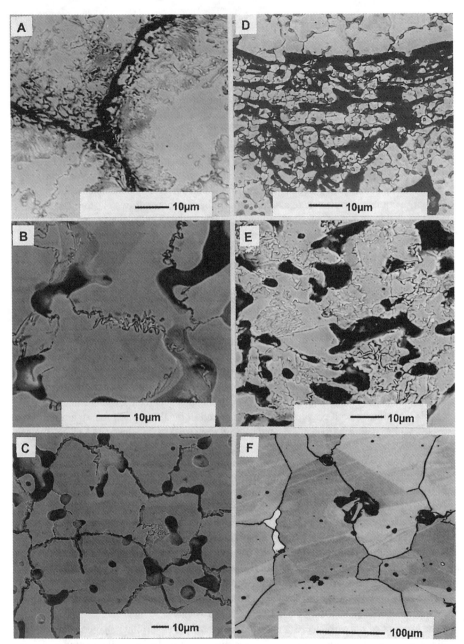

Figure 5 – *SEM Micrographs showing the differences in microstructure for molecular 75%H₂/25%N₂ sintered samples: (A) WA Coarse Porosity (B) WA Medium Porosity (C) WA Fine Porosity, (D) BM Coarse Porosity, (E) BM Medium Porosity (F) BM Fine Porosity. (WA = water atomised, BM = bimodal)*

The reasons for this behaviour could be related to either: 1) the movement of the sintering atmosphere through the sample, controlling the type and location of the nitride precipitation, or 2) dissociation of the discrete chromium nitride precipitates at higher sintering temperatures, the chromium returning into solution while the nitrogen was lost. At high sintering temperatures the most likely reason for nitrides disappearing from the microstructure was the result of individual grain growth; longer sintering times meant larger grains, forcing the chromium nitrides to remain at the grain boundaries. However, further work would be needed to confirm this.

Sintering in molecular $75\%H_2/25\%N_2$ appeared to encourage the formation of a liquid phase at slightly lower temperatures, possibly indicating that nitrogen lowered the solidus temperature. Consulting the ternary diagram for chromium-carbon-nitrogen system suggested that M_7C_3, could be replaced, in the presence of nitrogen, by $M_{23}C_6$ and a carbo-chromium-nitride structure, $Cr_2(CN)$. Therefore, the carbide eutectic networks in the molecular $75\%H_2/25\%N_2$ sintered bimodal samples could have been some form of carbo-chromium-nitrides, an apparently more brittle structure than the M_7C_3.

Mechanical Properties

Macro- and Micro-hardness – Micro-hardness values, taken from the within the grains, showed that sintering under vacuum gave rise to the lowest micro-hardness values of $260Hv_{10}$ whilst sintering in either a flowing molecular mixture of $75\%H_2/25\% N_2$ or in argon gave similar micro-hardness values of $\sim345Hv_{10}$.

Macro-hardness values increased as the density of the compact increased so that at near full densities, macro-hardness values were equivalent to the wrought baseline material. As with the micro-hardness values, different sintering atmospheres gave rise to variations in the macro-hardness results. The lowest values were recorded after vacuum sintering ($50\text{-}170Hv_{10}$) whilst both argon sintered samples and molecular $75\%H_2/25\% N_2$ sintered samples exhibited similar ranges of values ($68\text{-}250Hv_{10}$ and $68 \text{-} 200Hv_{10}$ respectively).

Tensile Testing – No water atomised samples were tensile tested because of the high compaction pressures required, which exceeded both the burst pressure of the rectangular die and the 50 ton load limit of the press which was the minimum load required to manufacture the coarse porosity samples. Therefore all tensile testing results refer to the bimodal powders only. Note that the compaction pressure of the fine porosity samples was dropped from 1080MPa to $\sim$ 900MPa because of the loading limitation.

In general, all the tensile properties such as ultimate tensile strength (UTS) and percentage elongation increased as the total porosity decreased, (Table 4). This was not unexpected and was in line with the literature on the mechanical properties of PM materials, where the lower the porosity the lower the interconnected porosity, so as the load bearing area was increased, the tensile properties improved. Although this trend was evident, variations in properties were also noted dependent on the sintering atmosphere.

Vacuum sintered bimodal powders exhibited lower UTS and yield strengths than either the argon or molecular $75\%H_2/25\% N_2$ sintered samples, with the latter samples exhibiting the highest tensile and yield strengths (Table 4), although the tensile strength of these samples was still 50% lower than the wrought value.

Fracture surfaces of the bimodal samples, and the results above, revealed that these materials had limited ductility. Fractographic studies indicated that for both the vacuum and argon sintered samples, the fine flaky ball milled particles failed in a ductile manner at the interparticle necks (Figure 6b), while the larger coarse water atomised powders remained intact leaving large spherical holes in the opposing fracture face, (Figure 6a), suggesting that fracture had occurred along prior particle boundaries. Both the coarse and medium porosity samples showed clear dimple sites associated with ductile fracture, the number and size of which increased as the compaction pressure and sintering temperature increased.

Table 4 – *Mechanical properties for bimodal powders sintered in three atmospheres*

Sample	Atmosphere	T_p (%)	T_{icp} (%)	UTS (MPa)	0.2% Yield Stress (MPa)	Elongation (%)	Reduction in area (%)
Coarse		29.4	28.0	217 ± 22	210 ± 26	0.15 ± 0.2	0.2 ± 0.04
Medium	Vacuum	24.0	18.5	261 ± 25	235 ± 36	0.4 ± 0.05	0.6 ± 0.2
Fine		20.1	13.0	296 ± 9	257 ± 17	1.7 ± 0.5	2.3 ± 0.5
Coarse		30.2	16.7	150 ± 4	147 ± 11	0.2 ± 0.04	0.1 ± 0.04
Medium	Argon	21.8	12.5	362 ± 5	322 ± 7	0.6 ± 0.06	0.3 ± 0.1
Fine		11.8	0.5	497 ± 10	456 ± 4	0.6 ± 0.15	0.3 ± 0.1
Coarse		30.2	23.1	217 ± 32	210 ± 16	0.4 ± 0.1	0.4 ± 0.2
Medium	75%H_2/25%N_2	24.4	23.9	440 ± 15	399 ± 9	0.9 ± 0.2	0.4 ± 0.2
Fine		4.8	2.08	635 ± 21	563 ± 4	1.4 ± 0.7	0.9 ± 0.5
MC - F75		-	-	1123 ± 18	948 ± 7	8.7 ± 1.0	9.7 ± 1.6

Note: MC-F75 = Modified Carbon ASTM F75 - fully dense, T_p = Total porosity & T_{icp} = total interconnected porosity

Molecular 75%H_2/25% N_2 sintered samples had, in effect, become sensitised at the grain boundaries, due to chromium nitride precipitation, so that some of the large water atomised particles had fractured along the grain boundaries where the large cellular/lamellar chromium nitride precipitates had formed.

Interestingly, both the fine porosity samples sintered in molecular 75%H_2/25% N_2 and the fine porosity samples sintered in argon fractured in a different manner to that seen with the other samples. With the argon sintered samples, non-continuous carbide eutectics were present at the fracture face suggesting that cracks had progressed along the grain boundaries and along the eutectic/grain boundary interface (Figure 6c).

A different situation occurred with the fine porosity samples sintered in molecular 75%H_2/25% N_2. In this case, as opposed to the microstructure presented in (Figure 5f), both large chromium nitride lamellar and carbide eutectics were present at the grain boundaries with the carbides sited inside the nitrides. Broken halves of these nitride and carbide clusters could be seen running across the majority of the fracture face while just below the fracture, cracks were seen in the carbide eutectics (Figure 6d). EDAX analysis of the fracture surfaces for these samples indicated higher levels of carbon and nitrogen compared to the fracture surfaces of the argon and vacuum sintered samples suggesting that the crack had progressed along grain boundaries due to the cleavage failure of either the nitrides or the carbides or a combination of the two.

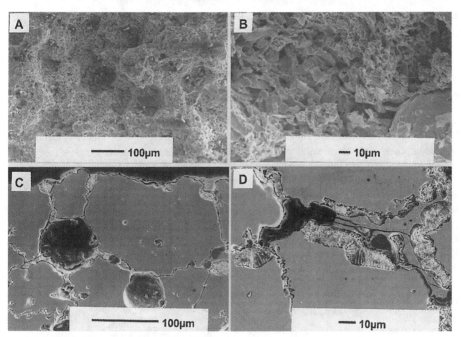

Figure 6 – *Fractographs of bimodal powder samples: (A) general view of fracture surface, (B) ductile failure of ball milled powder, (C) longitudinal section (electrolytically etched in HCl) of fine porosity sintered in argon, (D) longitudinal section (electrolytically etched in HCl) of fine porosity sintered in molecular 75%H$_2$/25%N$_2$*

Conclusions

1) Creating a bimodal powder from an equal blend of ball milled PREP powder and water atomised powder (353-106μm) gave similar porosity levels to the individual particle fractions of the water atomised powders but at lower compaction pressures and with better green strengths.

2) Variations in the initial powder morphology changed the packing characteristics and therefore changed the diffusion of the sintering atmosphere through the compact. This

altered the formation, size and type of the precipitate formed within one material type and the levels of surface oxide reduction. The extent of precipitate/carbide eutectic formation was also dependent on sintering temperature with the largest variations in density being obtained at sintering temperatures greater than 1250°C.

3) Sintering atmosphere had varying effects on the powders. Vacuum sintering led to high levels of chromium volatilisation and decarburisation. Sintering in argon proved to be a highly protective atmosphere permitting large volumetric quantities of M_7C_3 or $M_{23}C_6$ carbides to form. Whilst sintering in a molecular mixture of $75\%H_2/25\%N_2$ led to the formation of $Cr_2(CN)$ precipitates along with the M_7C_3 or $M_{23}C_6$ carbides.

4) Differences in microstructure also led to differences in mechanical properties, such as hardness, tensile strength, percentage elongation and Young's Modulus. Generally sintering in vacuum gave the lowest values, whilst sintering in argon or molecular $75\%H_2/25\%N_2$ gave the highest values. Slight solid solution strengthening of the matrix with nitrogen produced slightly higher tensile values and hardness, then with the same material sintered in argon.

5) By altering the processing parameters a range of porous materials were developed with both high total porosity and low total porosity. Although the mechanical properties of the low total porosity bimodal materials were substantially less than for the fully dense alloy tested in this work, tensile and yield strengths were not too dissimilar from conventional ASTM F75 alloys. Therefore, low total porosity bimodal materials should be able to withstand the physiological loads. Large pores, suitable for polymer impregnation, were developed in both the high porosity bimodal and high porosity water atomised materials. Another study suggested that the high porosity water atomised material would provide a better pore geometry for polymer impregnation. Thus a porosity graded cup could be developed by utilising the best characteristics of the two powders studied, providing that a suitable manufacturing process could be developed.

Acknowledgements

This work was effected within a Brite/EuRam project BE-4249-90 and therefore the co-operation of all the European Partners is acknowledged. The authors would like to thank Dr. Eric Jones, Howmedica International Inc., Limerick, Ireland for supplying the Co-Cr-Mo bar stock for atomisation and Dr. Manuela M. Oliveira, Instituto Nacional de Engneharia e Tecnolgia Industrial, Lisbon, Portugal for undertaking the first batch of Co-Cr-Mo water atomisation.

References

[1] Unsworth, A., "Tribology of Human and Artificial Joints," *Proceedings of the Institution of. Mechanical Engineers - Part H*, 1991, Vol. 205, pp. 163-172.
[2] Fuson, F.L., "Powdered Metal - Orthopaedic Implants" *International Conference on Hot Isostatic Pressing*, Dundee, Ill, 1978, pp.1-5.
[3] Bardos, D.I., "High Strength Co-Cr-Mo Alloy by Hot Isostatic Pressing of

Powder," *Biomaterial medical Devices and Artificial Organs*, 1979, Vol. 7, No. 1, pp. 73-80.

[4] Raman, R.V. and Rele, R.V., "A Novel Rapid Consolidation Powder Metallurgy Approach to Fabricate High Fatigue Strength and Corrosion Resistant Co-Mo-Cr-C Alloys for Demanding Applications," *P/M in Aerospace, Defence and Demanding Applications*, F.H.Froes, Ed., Metal Powder Industries Federation, Princeton, NJ, 1993, pp. 317-324.

[5] Hirschhorn, J. S. and Reynolds, J.T., " Powder Metallurgy Fabrication of Cobalt Alloy Surgical Implant Materials," *Research in Dental and Medical Materials*, E. Korostoff, Ed., Plenum Press, NY, 1969, pp.137-150

[6] Huddleston, J. B., "Powder Production Methods, Size and Powder Morphology and How it Affects the User," *Thermal Spray Research and Applications*, Metal Powder Industries Federation, Princeton, NJ, 1990, pp. 321-324.

[7] Pilliar, R. M., "Powder Metal-Made Orthopaedic Implants with Porous Surface for Fixation by Tissue In-growth," *Clinical Orthopaedics and Related Research*, 1983, No. 176, pp. 42-51.

[8] Pilliar, R. M., MacGregor, D. C, MacNab, I., Cameron, H.U., "P/M Surface Coatings on Surgical Implants," *Modern Developments in Powder Metallurgy*, H. H. Hausner and P. V. Taubenblat, Eds., Metal Powder Industries Federation, Princeton, NJ, Vol. 11, 1979, pp. 263-278.

[9] Miyamoto, Y., Tanihata, K., Matuzaki, Y., and Ma, X., "Synthesis of Functionally Gradient Materials by the Gas-Pressure Combustion Sintering," *The 35th Japan Congress on Materials Research*, Kyoto, Japan, 1992, pp. 1-6.

[10] Hahn, H., Lare, P. J., Rowe, R. H., Jr., Fraker, A. C., and Ordway, F., "Mechanical Properties and Structure of Ti-6Al-4V with Graded-Porosity Coatings Applied by Plasma Spraying for Use in Orthopedic Implants," *Corrosion and Degradation of Implant Materials: Second Symposium. ASTM STP 859*. A. C. Fraker, C. D. Griffin, Eds. American Society for Testing and Materials, Philadelphia, 1985, pp. 179-191.

[11] Anon, "Functionally Gradient Material Used in Artificial Teeth," *Biomedical Materials*, 1994, February, p. 4.

[12] Raghu, T., Malhotra, S. N., Ramakrishnan, P., "Corrosion behaviour of Porous Sintered Type 316L Austenitic Stainless Steel in 3% NaCl Solution" *Corrosion*, 1989, Vol. 45, No. 9, pp. 698-704.

[13] Kilner, T., Pilliar R. M., Weatherly, G. C., and Allibert, C., "Phase Identification and Incipient Melting in a Cast Co-Cr Surgical Implant Alloy," *Journal of Biomedical Materials Research*, 1982, Vol. 16, pp. 63-79.

[14] Becker, B. S., Bolton, J. D., and Youseffi, M., "Production of Porous Sintered Co-Cr-Mo Alloys for Possible Surgical Implant Applications - Part 1: Compaction, Sintering Behaviour and Properties," *Powder Metallurgy*, 1995, Vol. 38, No. 3, pp. 201-208.

[15] Kilner, T., Dempsey, A.J., Pilliar R. M. and Weatherly, G. C., "The Effects of Nitrogen Additions to a Cobalt-Chromium Surgical Implant Alloy," *Journal of Materials Science*, 1987, Vol. 22, pp. 565-574.

Richard M. Berlin,[1] Larry J. Gustavson,[1] and Kathy K. Wang[1]

Influence of Post Processing on the Mechanical Properties of Investment Cast and Wrought Co-Cr-Mo Alloys

---

Reference: Berlin, R. M., Gustavson, L. J. and Wang, K. K., "**Influence of Post Processing on the Mechanical Properties of Investment Cast and Wrought Co-Cr-Mo Alloys**", *Cobalt-Base Alloys for Biomedical Applications, ASTM STP 1365*, J. A. Disegi, R. L. Kennedy, and R. Pilliar, Eds., American Society for Testing and Materials, West Conshohocken, PA, 1999.

Abstract: The mechanical properties of investment cast Co-Cr-Mo alloy (American Society for Testing and Materials F 75) can be affected to varying degrees by post cast processes such as solution treating (ST), hot isostatic pressing (HIP), sintering used to apply porous coatings, repair welding, abrasive blasting, and laser marking. The mechanical properties of the wrought version of the alloy (American Society for Testing and Materials F 1537) can be influenced by mill practices. Thermo-mechanical processing such as forging, will change the properties of mill products depending on forging practices. Post forging processes such as abrasive blasting and laser marking can affect the mechanical properties to varying degrees.

Testing has shown that abrasive blasting has no significant effect on either alloy. Laser marking can reduce the fatigue strength of both alloys. Sintering the cast alloy will reduce the fatigue strength and that HIP will improve the fatigue strength of the sintered cast alloy. Also, the cast alloy can be repair welded with no loss in tensile properties.

Keywords: sintering, fatigue, hot isostatic pressing, cobalt-chromium-molybdenum alloy, investment casting, wrought material, porous coatings

Over the past sixty years cobalt-chromium-molybdenum alloy has demonstrated a remarkable level of versatility and durability as an orthopedic implant material. Originally introduced in the early 1930's by Austenal Laboratories (Howmedica) in the

---

[1] Principal Research Engineer, Director and Assistant Director, respectively, Research and Development, Howmedica Inc., NJ 07070.

cast form for dental applications, Vitallium®[2] alloy was soon adopted for orthopedic use owing to its exceptional corrosion resistance and biocompatability, compared with other material in use at the time. Since its introduction, cast Vitallium® alloy has been employed in virtually every articulating joint in the body, as well as, in numerous fracture fixation applications. In the 1970's the alloy was slightly modified to permit thermo-mechanical processing, which lead to the introduction of high strength forged cobalt-chromium-molybdenum femoral stems for total hips. Developed in response to reports of femoral hip stem fatigue failures, forged Vitallium® alloy provided nearly twice the fatigue strength of the cast alloy.

Investment casting and closed die forging remain the principal fabrication processes used to manufacture cobalt-chromium-molybdenum alloy implants. How these processes are performed can markedly affect the mechanical and metallurgical properties of the resulting cast and forged components. If not performed properly, the casting process can introduce material defects or microstructural features, which can reduce the tensile and fatigue strength. Forged material strength, hardness and microstructure can vary, depending upon the forging temperature and the nature and extent of material deformation imparted during forging. Other processing operations common to implant manufacturing, such as thermal processing, abrasive blasting, laser marking, etc. can also impact on the alloy's mechanical properties, both positively and negatively. In the sections to follow the impact of these processes on the properties of cobalt-chromium-molybdenum alloy will be discussed.

Materials and Test Methods

This paper evaluates the effects of processing on the investment cast American Society for Testing and Materials F 75 alloy with a nominal carbon content of 0.24 wt.% and the wrought version of the alloy (American Society for Testing and Materials F 1537) with a nominal carbon content of 0.05 wt.%. The wrought alloy was also evaluated in the forged condition per American Society for Testing and Materials F 799.

One of the fatigue samples used for this paper is a fatigue-bending rotating beam specimen which has a minimum diameter of 4.77 mm at the location of maximum stress. All fatigue samples were tested on a Fatigue Dynamics RBF 200 Fatigue Testing Machine which rotates at a maximum of 10,000 revolution per minute (167 Hz). The equipment is typically run in air at 70% of maximum output or at about 7000 revolutions per minute (117 Hz). The fatigue test was terminated if either the specimen fractured or the sample exceeded 10^7 cycles. This type of load configuration applies a uniform cantilever bending moment across the surface of the specimen, with the maximum stress occurring at the minimum cross section. Alternating cycles of tension and compression result in a mean stress of zero (fully reversed loading) and a R value of minus one (A = ∞). The test sample and test set up can be seen in Figure 1.

---

[2]Vitallium®- Trademark of Howmedica.

To test the effect of laser etching a second fatigue sample configuration was used. This was a four-point bend sample with a minimum gage section of 5.38 mm x 15.48 mm. All samples were tested in air on a Sonntag SF01U fatigue testing machine at a constant frequency of 30 Hz. The fatigue test was terminated if either the specimen fractured or the sample exceeded 10^7 cycles. The loading condition was set to produce a R value of 0.1. The test sample and test set up can be seen in Figure 2.

When tensile testing was performed, the cast samples were typically cast to size and the wrought or forged materials were machined into the appropriate tensile sample. All tensile samples and testing requirements met the requirements of American Society for Testing and Materials E 8.

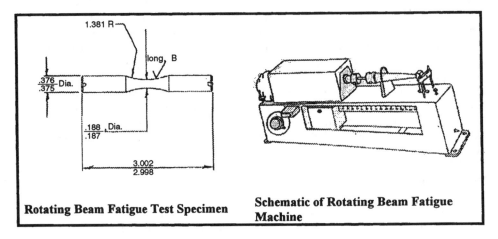

Rotating Beam Fatigue Test Specimen   **Schematic of Rotating Beam Fatigue Machine**

Figure 1-- *Rotating beam fatigue test sample and test set-up.*

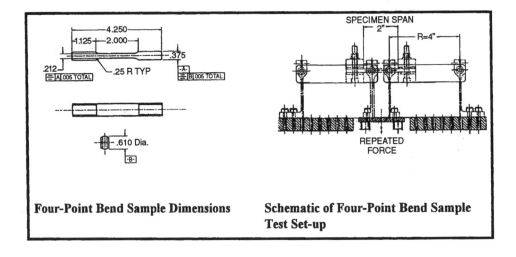

Four-Point Bend Sample Dimensions   **Schematic of Four-Point Bend Sample Test Set-up**

Figure 2-- *Four-point bend test sample and test set-up.*

Results and Discussion:

Post Casting Thermal Processes

Thermal processes, such as, solution or homogenization heat treating and hot isostatic pressing (HIP) are applied to cast ASTM F 75 alloy to enhance properties by increasing microstructural homogeneity and soundness. Sintering, a thermal process employed to create an interconnected porous coating on non-cemented implants surfaces tends to lower alloy mechanical properties. Table 1 presents typical room temperature tensile properties for cast ASTM F 75 alloy in the as cast, solution treated, sintered, and sintered and HIP conditions.

Table 1 - *Typical Tensile Properties of ASTM F 75 alloy [2]*

Condition	UTS (MPa)	0.2% YS (MPa)	El. (%)	RA (%)
As Cast	731	559	9.3	12.0
Solution Treated (ST)	842	531	17.3	16.5
Sintered[1]	787	524	12.3	11.9
Sintered + HIP + ST	752	508	16.4	13.6

[1] The sintering cycle is that used at Howmedica, other sintering processes may yield different results. This is true for all testing performed on sintered samples for this paper.

Typical Tensile Properties

The as cast alloy tends to provide a higher yield strength than is achieved following heat treating. But, heat treating at about 1220°C improves the ductility and ultimate tensile strength, while improving microstructural homogeneity. Sintering tends to lower all aspects of the tensile properties. HIP'ing and solution heat treating after sinter increases the ductility while improving the microstructure through healing of voids caused by exposure to the elevated temperature required for sintering.

Table 2 presents typical room temperature Rotating beam fatigue properties for cast ASTM F 75 alloy in the as cast, solution treated, sintered, and sintered and HIP conditions.

Table 2 – *Rotating Beam Fatigue Strength of ASTM F 75 Alloy [7]*

Heat Treat Condition	Surface Finish	Fatigue Strength at 10^7 Cycles (MPa)
As Cast	Smooth	345-480
HIP and Solution Treated (ST)	Smooth	380-450
Sintered	Smooth	210-275
Sintered + HIP + ST	Smooth	345-380

The as cast alloy can achieve the highest fatigue strength depending on the soundness of the casting. HIP is employed on cast alloys to close internal porosity, which inevitably exist in cast structures to varying degrees. A sound casting with minimal internal porosity would show a lesser effect from HIP'ing than would one with substantial porosity. Therefore the degree of property improvement of cast alloys due to HIP can be expected to be variable. For cast cobalt-chromium-molybdenum alloy the benefit of HIP is most consistently observed in the restoration of fatigue properties of cast alloy following exposure to the porous coating sinter cycle. The elevated temperature involved in porous coating sintering can result in as much as a 40% drop in rotating beam fatigue strength of the solution treated alloy. In these instances, HIP has been shown to reverse the property loss (Table 2) and restore the fatigue strength to close to that of the original solution treated alloy. HIP of cast cobalt-chromium-molybdenum alloy should always be followed by a heat treatment to resolution the carbides which generally precipitate during the slow cool which typically follows most HIP cycles.

Repair Welding

Repair welding is a process which, like casting, can yield variable results depending on how well it is executed. Cast cobalt-chromium-molybdenum alloy can be repair welded with no loss in tensile properties, provided the welding is performed properly. To demonstrate cast to size ASTM F 75 alloy tensile bars were machined to create a 90° notch at the midpoint of their gage section. The notch region was then repaired by tungsten inert gas welding using cast ASTM F 75 alloy filler rod. The weld repaired bars were then NDT examined using X-ray and FPI, solution treated and machined to match the original tensile bar dimensions. The bars were then tested in room temperature tensile and the results compared to properties of non welded cast and solution treated alloy (Table 3).

Table 3 – *Tensile Properties of ASTM F 75 Alloy with Repair Welding [2]*

Condition	UTS (MPa)	0.2% YS (MPa)	El. (%)
Repair Welded + Solution Treated	856	580	18.2
Solution Treated	842	531	17.3

Weld repairing (welding and solution treated), when done properly, will yield tensile results similar to that of solution treated alloy.

Thermo-mechanical Processing

For the wrought version of the Co-Cr-Mo alloy (ASTM F1537) there is very little fatigue data in the literature [1]. Tensile testing was performed on both mill products (annealed and warm worked, previously known as hot-rolled unannealed bar) and forged components (ASTM F799). Table 4 shows a comparison of the data. The minimum ASTM tensile properties required of the forgings is the same as that for the warm worked bar. The properties of warm worked bar has been shown to be dependent on the diameter of the mill product. The properties of the as forged material is dependent upon the forging temperature and the nature and extent of material deformation imparted during forging

Table 4 – *Tensile Properties of Wrought Co-Cr-Mo Alloy [2]*

Condition	UTS (MPa)	0.2% (MPa)	El. (%)	RA (%)
Warm Worked and Forged				
(ASTM Minimum)	1172	827	12	12
(Typical)	1332	1001	17	20
Mill Anneal				
(ASTM Minimum)	897	517	20	20
(Typical)	1001	994	26	23
Typical Forged[1]	1449	994	26	26

[1]Forged properties are typical of Howmedica forged products.

Table 5 presents typical rotating beam fatigue properties for the wrought cobalt-chromium-molybdenum alloy in various mill conditions and as forged. The warm worked bar and the as forged alloy exhibit similar tensile and fatigue properties, which are significantly higher than those of the annealed material.

Table 5 – *RBF Fatigue Results of Wrought and Forged Co-Cr-Mo Alloy [2]*

Condition	Surface Finish	Fatigue Strength at 10^7 Cycles (MPa)
Mill Anneal	Smooth	483
Warm Worked	Smooth	690
Forged	Smooth	759-828

Abrasive Blasting

Abrasive blasting is commonly employed to improve conditions for cement fixation on cemented devices. Blast surface finishes range from a smooth surface with an Ra in the range of 0.25 μm, to heavy grit blast surfaces with an Ra up to 10.16 μm. There are other blast finishes with intermediate Ra values that are also used. Rotating beam fatigue testing was performed on both the cast and wrought cobalt-chromium-molybdenum alloy bars having different blast surfaces to determine the sensitivity of the material to the final surface finish. Table 6 presents these results.

Table 6 - *Rotating Beam Fatigue Strength of Abrasive Blasted Co-Cr-Mo Alloy [2]*

Alloy Version	Condition	Surface Finish Ra (μm)	Fatigue Strength at 10^7 Cycles (MPa)
Cast	HIP + ST	0.25	380-449
Cast	HIP + ST	5.99	414-449
Cast	HIP + ST	10.16	414-449
Wrought	Warm Worked	10.16	656-690
Wrought	Forged	10.16	794

This testing shows that abrasively blasting to a surface roughness of up to Ra = 10.16 μm has little impact on the alloy fatigue strength of either the cast or wrought versions of the alloy.

Laser Marking

Laser marking is commonly employed as a means to identify or label metallic implants. To evaluate the effects of laser marking on the fatigue strength of cast and wrought Co-Cr-Mo alloy fatigue testing was performed using four point bend fatigue samples which were laser marked in the gage section. The four point bend specimen was chosen for the study because the stress is uniform over the specimen's gage length and therefore the location of the laser mark would not influence the potential for failure. Testing was performed on cast and wrought warm worked bar where the cast alloy samples were cast to size and the wrought were machined from warm worked bar. The laser marking was performed following all heat treating and finishing operations.

As can be seen from the data in Table 7, both the cast and wrought Co-Cr-Mo alloys exhibited a drop in fatigue strength due to laser marking. The cast alloy was most affected in the post sinter condition with a 36% loss, while the wrought alloy experienced over a 60% drop in fatigue strength. In all instances, for the laser marked samples, fracture initiated in the laser marking.

Table 7-- *Four Point Bend Fatigue Strength of Laser Marked Co-Cr-Mo Alloys [2]*

Alloy Version	Condition	Laser Marked	Fatigue Strength at 10^7 Cycles (MPa)
Cast	HIP + ST	No	449-483
Cast	HIP + ST	Yes	345-380
Cast	Sinter + HIP + ST (thermal cycle only)	No	380
Cast	Sinter + HIP + ST (thermal cycle only)	Yes	242
Wrought	Warm worked bar	No	> 828
Wrought	Warm worked bar	Yes	310

Conclusions

For the casting alloy (American Society for Testing and Materials F 75) it was found that the cast alloy is capable of much higher fatigue properties than the previously published literature [3-7] presented. Repair welding of castings, when done correctly, can yield equivalent tensile properties to the cast and non-welded material.   That abrasively blasting to a surface roughness of up to Ra = 10.16 µm has little impact on the alloy fatigue strength. Laser marking can reduce the fatigue strength by up to 36%.

For the wrought and forged alloy (ASTM F 1537 and F 799) it was found that the fatigue strength of the warm worked bar and as forged material are much higher than the annealed bar. That like the cast alloy abrasively blasting to a surface roughness of Ra = 10.16 µm has little impact on the fatigue strength of the high strength wrought of forged alloy. Laser marking can reduce the fatigue strength of the high strength wrought alloy up to 60%.

References:

[1] Kumar, Prabhat,. Hickl, Anthony J, Asphahani, Aziz I. and Lawley, Alan, *"Properties and Characteristics of Cast, Wrought and Powder Metallurgy (P/M) Processed Cobalt-Chromium-Molybdenum Implant Materials"*, American Society for Testing and Materials *STP 859*, pp 30-56.

[2] Internal Howmedica reports, Data on file at Howmedica R&D.

[3]. Georgette, Frederick S, *"Effect of Hot Isostatic Pressing on the Mechanical and*

Corrosion Properties of a Cast, Porous-Coated Co-Cr-Mo Alloy", American Society for Testing and Materials *STP 953*, pp 31-46.

[4] Georgette, F.S., and Davidson, J.A., "*The effect of HIP'ing on the fatigue and tensile strength of a cast, porous coated Co-Cr-Mo alloy*, Journal of Biomaterials Research, 1986, Vol. 20 pp 1229-1248

[5] Spires, Jr., Walter P., Kelman, David C., and Pafford, John A., "*Mechanical Evaluation of ASTM F 75 Alloy in Various Metallurgical Conditions*", American Society for Testing and Materials *STP 953*, pp 47-59.

[6] Pilliar, Ph.D., R.M. , "*Powder Metal-Made Orthopedic Implants with Porous Surface for Fixation by Tissue Ingrowth*", Clinical Orthopedics and Related Research, June 1983 Volume 176, pp 42 - 51.

[7] Gustavson, L., Crippen, T., Dumbleton, J.H.and Bushelow, M., "*Cast Co-Cr-Mo Alloy as a Substrate for Porous Coated Prosthetic Devices*", Materials Research Society Symposium Proceedings Volume 110, Biomedical Materials and Devices, pp 553-559.

Ajit K. Mishra[1], Mark A. Hamby[1], and William B. Kaiser[1]

Metallurgy, Microstructure, Chemistry and Mechanical Properties of a New Grade of Cobalt-Chromium Alloy Before and After Porous-Coating

Reference: Mishra, A. K., Hamby, M. A., and Kaiser, W. B., "**Metallurgy, Microstructure, Chemistry and Mechanical Properties of a New Grade of Cobalt-Chromium Alloy Before and After Porous-Coating,**" *Cobalt-Base Alloys for Biomedical Applications, ASTM STP 1365*, J. A. Disegi, R. L. Kennedy, and R. Pilliar, Eds., American Society for Testing and Materials, West Conshohocken, PA, 1999.

Abstract: Titanium alloy (Ti-6Al-4V) and cast cobalt-chromium-molybdenum alloy (CoCr) have a long history of successful clinical use. However, a higher strength material would be useful to manufacture certain implants, such as smaller hip-stems, with more extensive porous-coating to facilitate improved fixation. A new grade [high carbon (C), proposed grade 2] of CoCr, within the ASTM specification for wrought cobalt - 28 chromium - 6 molybdenum alloy for surgical implants (F 1537), has been developed to meet this objective. One producer's version of this new material was compared with cast CoCr. The typical C content of this material is similar to that of cast CoCr. In the mock-sintered condition, the tensile and yield strength of bar and forgings of this new CoCr were similar to each other and greater than that of cast CoCr, and all three materials had similar ductility. In both porous-coated and mock-sintered conditions, the average grain size of bar and forgings of this new CoCr was 63 ± 19 to 145 ± 12 μm, while that of two commercially available porous-coated cast CoCr hip-stems from two manufacturers was 2540 ± 395 and 3394 ± 876 μm respectively. Metallography revealed that the cast CoCr implants contain the eutectic phase [face centered cubic (fcc) Co, the intermetallic sigma (σ) phase, and metal carbides ($M_{23}C_6$ and M_7C_3)] in the interdendritic regions. The 10 million cycle fatigue strength was 207 MPa for cast CoCr and 241 MPa for the new CoCr bar in the porous-coated condition, and 345 MPa for cast CoCr and at least 448 MPa for the new CoCr bar in the mock-sintered condition. Fatigue testing was performed only on bar since the tensile properties and microstructure of bar and forgings were very similar, both in the porous-coated and the mock-sintered condition. Thus, the new grade of wrought CoCr exhibited higher fatigue strength than cast CoCr in both porous-coated and mock-sintered conditions. This may be attributed to the finer grain size of this material.

Keywords: Cobalt-chromium alloy, high carbon, bar-stock, forging, porous-coated, fatigue strength, tensile strength, microstructure

[1]Senior Research Projects Manager, Research Engineer I, and Research Engineer II respectively, Smith and Nephew Inc., Orthopaedic Division, 1450 Brooks Road, Memphis, TN 38116, USA.

Introduction

Cobalt-base alloys are generally used in applications which require wear resistance, corrosion resistance and/or thermal resistance. Many of the current commercial cobalt-base alloys are derived from the work of Elwood Haynes at the turn of the century [1]. He discovered the strengthening effect and corrosion resistance imparted to Co by Cr, and patented the CoCr alloys in 1907. Haynes named these alloys the Stellite alloys. He subsequently discovered the strengthening effects of molybdenum (Mo) and tungsten (W) in CoCr, and patented the ternary CoCrMo and CoCrW alloys; these were the first cobalt-base superalloys. The harder compositions of these alloys are used in cutting tools and other wear-type applications such as plow shares, oil well drilling bits, dredging cutters, hot trimming dies and internal combustion engine valves and valve seats. The softer and tougher compositions are used for high temperature applications such as gas-turbine vanes and buckets.

The first use of a cobalt-alloy in implant applications was as an investment cast dental alloy in 1936 [2]. This alloy is currently described in the ASTM specification for cast cobalt-chromium-molybdenum alloy for surgical implant applications (F 75) and is still widely used today in many applications such as the femoral component of knee prostheses and the humeral component of shoulder prostheses. Although this alloy is used solely in implant applications, a close analogue of this alloy called Stellite 21 was used initially in aircraft turbocharger blades and is still used as an alloy for wear resistance.

The ASTM F 75 CoCrMo alloy was subsequently modified to make it forgeable and this led to the development of the ASTM specification for cobalt - 28 chromium - 6 molybdenum alloy forgings for surgical implants (F 799). This alloy is available in the form of mill-products such as bar-stock which is used either for direct machining of a device (such as the femoral head of hip prostheses) or forging of a device (such as cemented hip-stems). Prior to 1994, both bar-stock and forgings used to be covered under ASTM F 799, which was split in 1994-95 into F 799 for forgings and F 1537 for bar-stock.

Currently used CoCr forgings and bar-stock (wrought CoCr) have excellent strength in the non porous-coated condition. CoCr forgings are used, for instance, in cemented hip-stems, and perform very well. However, this alloy contains a low level of carbon (typically about 0.07%) compared to cast CoCr (ASTM F 75) (typically about 0.25%). Hence, although wrought CoCr is stronger than cast CoCr in the non porous-coated condition, when it is porous-coated, the high temperatures associated with sintering cause greater grain growth in wrought CoCr, while the carbides and interstitial carbon in cast CoCr tend to pin the grain boundaries, thus reducing grain growth and producing higher strength than wrought CoCr in the porous-coated condition. This leads to greater reduction in yield strength during the sintering process, which is likely to translate to a greater reduction in fatigue strength as well. That is the reason why porous-coated CoCr devices have until recently been exclusively made of cast CoCr.

However, castings are inherently associated with large initial grain size, non-homogeneities and other casting defects, and also have lower yield strength than their wrought counterparts at equivalent ductility levels. If a wrought CoCr had a high carbon content similar to that of cast CoCr, it would be less susceptible to grain growth than low carbon wrought CoCr during porous-coating, and would be more homogeneous and have a higher yield strength and smaller initial grain size than cast CoCr, thus potentially achieving higher fatigue strength than both cast CoCr and low carbon wrought CoCr in the porous-coated condition.

The current paper describes the evaluation of such a CoCr alloy. The objective of this work was to identify and evaluate a CoCr alloy which has high fatigue strength in the porous-coated condition, for potential use in extensively porous-coated hip systems, especially the smaller sizes which may be subject to higher stresses. This would enable an increase in the extent of porous-coating coverage, compared to what is possible in Ti-6Al-4V hip-stems.

Materials

Four high carbon wrought CoCr alloys were identified as potential candidates for this application: (1) BioDur™ CCM Plus™ (CCM+) (Carpenter Technology, Reading, PA), (2) Hi-Carb TJA-1537™ (Allvac, Monroe, NC), (3) Supermet high carbon CoCr (Firth Rixson Superalloys, Derbyshire, England), (4) Micro-Melt® F 75 CoCr (Carpenter Technology, Reading, PA). All four are commercially available as wrought bar-stock.

One significant difference between the alloys is that the billets for the two Carpenter alloys are produced using powder metallurgy techniques (vacuum induction melting, followed by inert gas atomization and hot isostatic pressing) while the billets for the Allvac and Firth Rixson alloys are produced using conventional melting and thermomechanical processing techniques. The fabrication of bar-stock from these billets is performed using conventional thermomechanical processing techniques for all four materials.

The testing described in this paper pertains to the CCM+ alloy.

Metallurgy

Cobalt undergoes an allotropic phase transformation from a high temperature fcc phase to the low temperature hexagonal close packed (hcp) epsilon (ε)-Co when cooled, at about 422°C. This fcc phase is referred to as alpha (α) by some authors and gamma (γ)-austenite by others. Alloying elements such as iron (Fe), manganese (Mn), nickel (Ni) and C tend to stabilize the fcc structure while Cr, Mo, W and silicon (Si) tend to stabilize the hcp structure [3]. However, the fcc to hcp transformation is relatively sluggish even for pure cobalt. Attempts have been made to exploit this transformation to develop age-hardenable Co_3Ti and cobalt-tantalum (Co_3Ta) type precipitates; however, these alloys have not been commercialized.

The more important means of strengthening of cobalt alloys are solid solution strengthening and precipitation of carbides, especially $M_{23}C_6$ carbides. The role of various alloying elements in CoCrMo alloys is given in Table 1. Cr improves the strength and corrosion resistance of Co. Mo and C are also strengtheners and Ni helps improve forgeability (Table 1).

Table 1 - *Role of various alloying elements in CoCr alloys [3]*

Element	Effect
Cr	Improves oxidation and corrosion resistance, strengthens by formation of M_7C_3 and $M_{23}C_6$ carbides
Mo	Solid solution strengthener, also strengthens by formation of Co_3M intermetallic compound and M_6C carbide
Ni	Stabilizes fcc form of matrix, improves forgeability
C	Strengthens by formation of MC, M_7C_3, $M_{23}C_6$ and possibly M_6C carbides

Most commercial cobalt-alloys have a metastable, continuous fcc phase matrix at room temperature and may contain one or more types of carbides for strengthening [4].

The higher level of C in the high C, wrought CoCr alloys described above, compared to low C, wrought CoCr, would be expected to produce a slight improvement in the strength of the as-received material. However, the more significant improvement would be expected to be in the porous-coated condition, for the reasons described above.

Test Methods

Chemical Analysis

Chemical analysis was performed by Carpenter per the ASTM Test Methods for Chemical Analysis of High Temperature Electrical, Magnetic, and Other Similar Iron, Nickel and Cobalt Alloys (E 354) on each heat of CCM+, and the results were provided with the certification of the bar-stock. The chemistry of all 9 heats of CCM+ for which certification has been received from Carpenter to date was compiled. Every lot of CCM+ bar-stock which was tested as a part of this work was produced from one of these 9 heats.

Forgeability Studies

Forgeability studies were performed at BI Jet (Lansing, MI), Komtek (Worcester, MA) and Teledyne Portland Forge (Lebanon, KY) to determine whether the CCM+ alloy can be forged to the dimensional tolerances required in a hip-stem and achieve the mechanical properties specified in ASTM F 799, since the forgeability of this material was expected to be lower than that of low carbon wrought CoCr, due to its higher carbon content.

Porous-Coating Studies

CCM+ test specimens and hip-stems were porous-coated by BI Thortex (Clackamas, OR) and AstroMet Inc. (Cincinnati, OH), using 250-350 μm (-45,+60 mesh) ASTM F 75 CoCr beads, in several different process lots in order to identify an optimum porous-coating cycle and verify reproducibility of the process. Sintering was performed on coupons machined from as-received, unannealed CCM+ bar stock, as well as on forged CCM+ hip-stems.

Microstructure

Optical microscopy was performed on the following specimens using a Reichert-Jung metallograph (Reichert-Jung, Austria, Model # 38101, Serial # 389913), a Nikon Epiphot metallograph (Nikon, Enfield, CT), and a stereomicroscope (Carl Zeiss, Oberkochen, Germany, Model: SV8).

- As-received, unannealed CCM+ bar-stock
- As-received, low carbon wrought CoCr bar-stock
- As-received cast CoCr
- Hip-stem forgings produced from CCM+ bar-stock
- Two commercially available cast + sintered (porous-coated) + HIP'ed (hot isostatic pressed) + solution-treated CoCr hip-stems from two implant manufacturers
- Hip-stem forgings produced by BI Jet from CCM+ bar-stock, and subsequently mock-sintered/porous-coated and solution-treated by Thortex or AstroMet

  The mock-sintering treatment has a thermal cycle identical to that of the sintering cycle, however the porous-coating is not applied, so as to simulate the treatment received by the neck and other uncoated regions of a hip implant.

  Note that the HIP'ing cycle, which is used for cast CoCr to reduce the incidence of casting porosity, was not used for CCM+, since this is a wrought material and HIP'ing would not be expected to provide any benefit.
- Coupons machined from CCM+ bar-stock, sintered and solution-treated at Thortex or AstroMet

All coupons were sectioned using an abrasive disc saw and compression mounted in thermosetting epoxy resin. The specimens were wet ground using SiC papers through 2400 grit followed by final polishing for about 4-5 minutes using 0.05 μm silica suspension. After the specimens were polished and cleaned, they were heated to about 180°C and etched by swabbing with a mixture of 5 ml water, 60 ml hydrochloric acid and 6 g Cupric chloride.

The grain size of the specimens was determined using the Hillard circular intercept method per ASTM Test Methods for Determining Grain Size (E 112).

Tensile Testing

Tensile testing was performed per the ASTM Test Methods for Tension Testing of Metallic Materials (E 8), using the round specimen (Figure 8 in ASTM E 8) with a gage diameter of 6.35 mm, and the ASTM Test Method for Young's Modulus, Tangent Modulus and Chord Modulus (E 111) at Koon-Hall-Adrian (Portland, OR) on multiple lots of as-received CCM+ bar-stock, forged CCM+ hip-stems, mock-sintered+solution-treated CCM+ bar-stock, and forged+mock-sintered+solution-treated CCM+ hip-stems. The specimens were tested to yield at a constant rate of 0.127 mm/min. After yield, the speed was increased to achieve failure in about 1 additional minute. The load at failure divided by the original cross-sectional area was reported as the ultimate tensile strength

(UTS). The stress at the 0.2% offset in strain from the linear (elastic) region of the stress-strain curve was reported as the yield strength (YS). The % change in length and cross-sectional area at failure were reported as the elongation (Δl) and reduction of area (RA) respectively. Testing was performed in air at room temperature.

Porous-Coating Characterization

The average pore size and mean volume percent of porosity of the porous-coating was determined by the linear intercept method (point count method), at 100X magnification on an in-plane view, after infiltrating the coating with epoxy followed by grinding and polishing to mid-thickness of the coating. Thickness of the porous-coating was obtained by subtracting the measured substrate thickness from the measured total thickness of each coupon. Measurements were performed using a digital caliper with a repeatable resolution of 0.025 mm.

The sintered bead porous-coating was evaluated by static tensile pull-off strength as well as static lap shear strength testing. These tests were performed in accordance with the ASTM Test Method for Tension Testing of Porous Metal Coatings (F 1147) and ASTM Test Method for Shear Testing of Porous Metal Coatings (F 1044). FM-1000 adhesive was used for gluing the porous-coated surfaces to an opposing coupon or disc made from stainless steel. Tensile and shear test coupons were joined using 0.19 MPa gluing pressure applied by spring force and 2.5 hours of curing time at 177 °C, followed by air-cooling overnight. Lap shear and tensile pull-off tests were performed using a United FM20 testing machine.

Fatigue Testing

Fatigue test specimens were machined by Low Stress Grind Inc. (Cincinnati, OH). All mock-sintered specimens were sintered as 1.6 cm diameter blanks cut from CCM+ bar-stock, prior to final machining. Porous-coated specimens were machined to the final dimension in the gage section followed by application of the porous-coating to the central 2.5 cm of the gage length, and subsequent sintering. After sintering, the specimen grip shanks were finish ground to within 0.025 mm concentricity to minimize any bending stresses. The specimen gage diameter was 5.08 mm.

The 10 million cycle fatigue endurance limit, for both porous-coated and mock-sintered specimens, was determined by axial fatigue testing per the ASTM Practice for Conducting Force Controlled Constant Amplitude Axial Fatigue Tests of Metallic Materials (E 466) at 60 Hz and R = 0.1. Testing was performed in air at room temperature. Although many surgical implants are subjected to bending fatigue loading in-vivo, axial fatigue testing was performed since an ASTM standard does not currently exist for bending fatigue testing. Further, cracks initiate at the tensile surface of a specimen or device subjected to bending fatigue stresses, as they do during axial fatigue testing in tension.

Testing was initiated at a high stress level and, if failure occurred prior to 10 million cycles, the stress was reduced by 34.5 MPa for the next test. If a specimen ran-out to 10 million cycles without failure at a given stress level, subsequent tests were performed at the same stress level until 5 runouts were obtained at that stress level or a specimen failed prior to 10 million cycles. If a failure occurred, the stress level was reduced by 34.5 MPa again. Following completion of 5 runouts (with no failures) at a single stress level remaining specimens were used to complete the stress vs. no of cycles (S/N) curve by testing at higher stress levels.

Results

Chemical Analysis

The chemistry of all 9 heats of CCM+ for which certification has been received from Carpenter to date is given in Table 2. The average chemistry of the alloy was: Co-29.2Cr-6.2Mo-0.22Ni-0.63Mn-0.40Fe-0.57Si-0.24C-0.18N. All 9 heats meet the chemistry requirements of ASTM F 1537 and F 799 and draft ISO standard 'Implants for Surgery - Metallic Materials - Part 12: Wrought cobalt-chromium-molybdenum alloy' (5832-12). The C content of this material is within the range specified for the proposed new grade (grade 2, high C) of ASTM F 1537 wrought CoCr, higher than that of grade 1 (low C) wrought CoCr and similar to that of cast CoCr.

Table 2 - *Chemistry (weight %) of various heats of the CCM+ alloy*

Element	Heat 182339	Heat 182376	Heat 182377	Heat 182378	Heat 182381	Heat 182382	Heat 182383	Heat 182384	Heat 182417	Avg.	Std Dev
Cr	29.56	28.53	28.4	28.5	29.5	29.6	29.7	29.7	29.66	29.24	0.58
Mo	6.32	6.03	6.02	6.02	6.4	6.46	6.41	6.3	6.33	6.25	0.18
Ni	0.36	0.14	0.1	0.1	0.2	0.3	0.2	0.1	0.46	0.22	0.13
Mn	0.74	0.44	0.4	0.4	0.8	0.7	0.8	0.7	0.72	0.63	0.17
Fe	0.48	0.33	0.27	0.28	0.46	0.57	0.5	0.35	0.35	0.40	0.11
Si	0.73	0.39	0.3	0.3	0.7	0.7	0.7	0.6	0.7	0.57	0.18
C	0.25	0.24	0.24	0.24	0.24	0.25	0.25	0.24	0.25	0.24	0.01
N	0.17	0.17	0.17	0.18	0.17	0.19	0.18	NR	0.17	0.18	0.01
Co	61.45	63.94	64.21	64.21	61.8	61.48	61.6	NR	61.79	62.56	1.30

Forgeability Studies

All three forging vendors successfully produced near-net shape hip-stem forgings from CCM+ bar-stock. The initial attempt by one of the forging vendors failed inspection due to tears produced during removal of flash, however this was rectified in subsequent lots. The results of this forgeability study indicate that the forging vendors do have a learning curve associated with processing the CCM+ alloy; however they can perform the forging operations after they have gained some experience with this alloy.

Carpenter recommends that hot-working operations such as forging be performed after preheating at 1121 or 1149 °C [5]. Komtek forged the test pieces after preheating at 1149°C. The forging temperatures used by Jet and TPF have not been disclosed to the authors. Recent lots of forgings produced at Jet have been annealed for 30 minutes at

1093°C to optimize ductility. The annealing process increases ductility, although with an accompanying decrease in static strength, by dissolving some of the carbides.

Carpenter's data sheet [5] indicates that significant loss of ductility occurs in the CCM+ alloy from even small amounts of cold work; hence this has not been attempted. Cobalt alloys intended for cold working processes, to produce sheet, tube, wire etc., usually contain no more than 0.15% C [3].

Porous-Coating Studies

All lots of CCM+ received to date have been successfully porous-coated by both porous-coating vendors: AstroMet and Thortex. Thortex indicated that they have had to adjust the sintering temperature depending upon the chemistry of each lot of material; AstroMet indicated that they were able to porous-coat material from all heats using the same sintering cycle.

As shown in Figure 1, even small variations in the chemistry affect both the liquidus temperature (i.e. the melting point, the temperature at which the alloy turns completely into a liquid) as well as the solidus temperature (i.e. the incipient melting temperature, the temperature at which the first drops of liquid begin to appear) in most alloys. The sintering process is normally performed at a level between the liquidus and solidus temperatures and the position of the liquidus and solidus lines determines, from the lever rule, the amount of liquid that is available to accomplish bonding of the beads to the substrate (Figure 1).

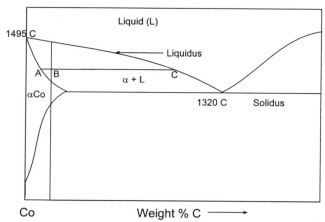

Figure 1 - *Schematic of Cobalt-Carbon phase diagram. Note that the lever rule states that at the sintering temperature AC, the fraction of the constituent phases which is liquid in an alloy of composition B is given by AB/AC.*

Hence, chemistry variations can potentially affect the sintering response of an alloy. The effect of various alloying elements on the liquidus temperature of Cobalt is given in Table 3 [6]. Comparing Tables 2 and 3, it is seen that the elements which are present in significant quantities in CCM+ and have the greatest effect on the sintering response are C, Si and Mn. In a binary Co-C alloy, increase in C% would decrease the melting point of the alloy from 1495 °C, the melting temperature of pure Co (Figure 1). Increase in C% would also cause an increase in the amount of liquid and decrease in the amount of solid αCo present at a given sintering temperature assuming that the sintering is being performed in the (α + liquid) phase region.

Table 3 - *Effect of 1 wt% of alloying element on melting temperature of Cobalt. Alloying elements are listed in the top row and the change in melting temperature (°C) is listed in the bottom row.*

W	N	Fe	Cr	Mo	V	Mn	Al	Ta	Zr	S	Ti	Nb	Si	B	C
+0.5	-0.5	-0.5	-3	-4	-8	-8	-11	-17	-17	-22	-36	-39	-42	-64	-67

It is likely that the best properties of the substrate in the porous-coated condition would be obtained if the sintering temperature can be minimized, with the consequent reduction in grain growth. Hence, the best material chemistry from this perspective is a high level of C, Si and Mn. However, high levels of C tend to reduce forgeability and high levels of Si can promote formation of intermetallic compounds such as σ, μ and Laves phases which can cause embrittlement [7]; hence, an appropriate balance has to be reached.

As described previously, the presence of carbides and interstitial carbon facilitates grain boundary pinning, and consequent reduction in grain growth, in the solid phase present during the sintering process and during the heat-up and cool-down steps before and after sintering.

Microstructure

An optical micrograph of unannealed CCM+ bar-stock is given in Figure 2. Note the fine grain structure. Fine (<10 μm), homogeneously distributed carbides are also observed. Optical micrographs of low C CoCr bar-stock and as-cast cast CoCr are given in Figures 3 and 4 for comparison. Note the single phase structure and absence of carbides in the former, and the larger grains and presence of carbides in the latter, as expected. A micrograph of a CCM+ forging is given in Figure 5. Note that the grain size of the forgings is larger than that of CCM+ bar-stock, but significantly smaller than that of cast CoCr.

Figures 6 and 7 show micrographs of two commercially available porous-coated cast CoCr hip-stems from two different manufacturers. The average grain size of the hip-stem from manufacturer A was 2540±395 μm, and that of the hip-stem from manufacturer B was 3394±876 μm. Note that the cast+sintered+HIP'ed+solution-treated implants from both companies exhibit large grains and significant grain size variability, and reveal the presence of carbides and the brittle eutectic phase at the grain boundaries and the interdendritic regions [8]. The presence of grain boundary carbides is expected since the grain boundaries provide sites for preferential precipitation of carbides. While the solution-treatment cycle reduces the incidence of grain boundary carbides through redissolution in the matrix, it can not eliminate this phenomenon entirely. There is no evidence of continuous grain boundary carbides, a microstructure which would be unacceptable due to its severe adverse effect on the ductility of the material.

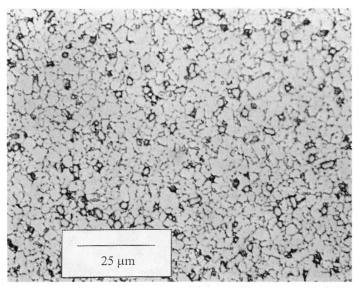

Figure 2 - *Optical micrograph of unannealed CCM+ bar-stock*

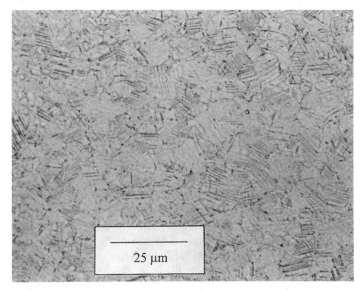

Figure 3 - *Optical micrograph of low carbon CoCr bar-stock*

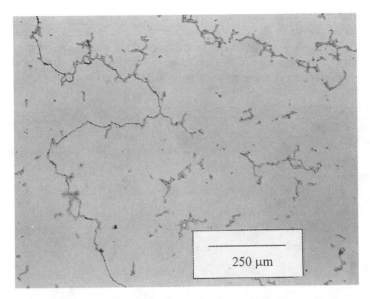

Figure 4 - *Optical micrograph of cast CoCr (as-cast)*

Figure 5 - *Optical micrograph of hip-stem forging produced from CCM+ bar-stock*

The interdendritic regions have a different chemistry (enriched in Cr and Mo) compared to the dendrites due to the phenomenon of 'coring' (segregation of the constituent elements depending upon their melting point) which is normal in castings, and hence they solidify at a lower temperature compared to the dendrites [8,9]. This results in segregation and solidification of the eutectic liquid in the interdendritic regions. The eutectic phase consists of fcc Co, the intermetallic σ phase, and $M_{23}C_6$ and M_7C_3 carbides [10,11]. The σ phase and the carbides have low ductility and are detrimental to the fatigue strength of the material. While the incidence of this interdendritic phase can be reduced by the solution-treatment cycle, it is difficult to eliminate it entirely in cast CoCr.

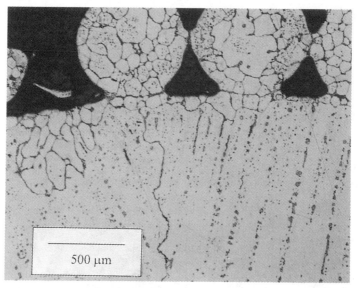

Figure 6 - *Optical micrograph of a commercially available porous-coated cast CoCr hip-stem from manufacturer A*

Figure 8 shows a micrograph of a hip-stem forging, produced by Jet from CCM+ bar-stock, and subsequently sintered and solution-treated by Thortex. The average grain size of 14 such devices (some processed at Thortex and some at AstroMet) was measured, and the results ranged from 63±19 to 145±12 μm, two orders of magnitude smaller than that of the cast+sintered+HIP'ed+solution-treated devices in Figures 6 and 7.

As in the cast+sintered+HIP'ed+solution-treated devices, the sintered CCM+ specimens (Figure 8) also show an incidence of carbides at grain boundaries. This is expected since both materials have similar carbon levels, and have been exposed to a high temperature sintering cycle, and the grain boundaries provide sites for preferential precipitation of carbides in both materials. There are isolated instances of grain boundary porosity in some areas. This is expected and is attributed to dissolution of some of the grain boundary carbides during the sintering and solution-treatment processes [10,12], consequent localized enrichment in C, which leads to reduction in the melting point, and localized incipient melting [12]. Note that there is no evidence of continuous grain boundary carbides, a microstructure which would be unacceptable due to its adverse effect on the ductility of the material.

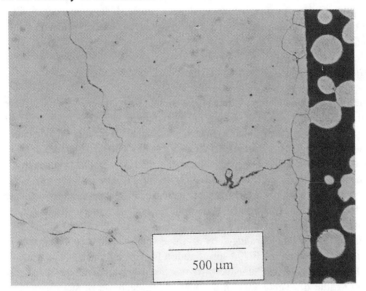

Figure 7 - *Optical micrograph of a commercially available porous-coated cast CoCr hip-stem from manufacturer B*

Figure 8 - *Optical micrograph of forged + sintered + solution-treated CCM+ hip-stem*

The average grain size of several lots of CCM+ bar-stock, sintered and solution-treated at Thortex or AstroMet, was also measured, and the results ranged from 37±12 μm to 124±4 μm, similar to that of forged+mock-sintered+solution-treated CCM+ (Figure 8), and two orders of magnitude finer than in the cast+sintered+HIP'ed+solution-treated devices (Figures 6 and 7).

Tensile Properties

Static tensile properties of 9 lots of CCM+ bar-stock (15 tests) are given in Table 4. As shown in the Table, the material meets the requirements of ASTM F 1537. The corresponding values for cast CoCr and low C, wrought Co-Cr are also given for comparison. The tensile and yield strength (UTS and YS) of CCM+ bar-stock are 50-80% greater than those of cast CoCr, as expected, and the ductility of the two materials is similar. All values are similar for CCM+ and low C, wrought CoCr.

The ductility of hot-worked CCM+ can be increased, if necessary, by annealing. However, this does result in reduction in UTS and YS and increase in grain size. The typical effects of annealing are shown in Table 5.

Tensile testing of the forgings produced by all three vendors revealed acceptable UTS and YS but lower ductility than the F 799 specification for CoCr forgings (Table 6). After annealing for 30 minutes at 1093 °C, the ductility increased to a level above that specified by ASTM F 799 (Table 6). There was an accompanying decrease in static strength, as expected; however the strength was still greater than specified in F 799.

Table 4 - *Tensile properties, hardness and grain size of wrought and cast CoCr*

	No. of Tests	UTS, MPa	YS, MPa	Δl, %	RA, %	Hardness, Rc	ASTM Grain Size
CCM+ bar-stock	15	1353±47	963±66	22±4	18±2	42±2	12±0
Cast CoCr	6	858±73	543±52	16±3	17±1	NT	NT
Low C CoCr bar	8	1322±59	900±40	26±3	23±3	42±4	10±0
ASTM F 1537 spec		1172	827	12	12	35	5

Table 5 - *Effect of annealing on typical tensile properties of CCM+ [5]*

Condition	UTS, MPa	YS, MPa	Δl, %	RA, %	Hardness, Rc	ASTM Grain Size
Hot-worked	1365	931	22	17	43	12
Annealed (1093 °C /1hr +WQ)	1351	883	22	18	42	10-11
Annealed (1121 °C/1hr +WQ)	1400	848	28	24	41	10-11
Annealed (1149 °C/1hr +WQ)	1393	807	29	25	40	10-11
Annealed (1204 °C/2hr +WQ)	1310	717	32	27	36	NS

The results of the static tensile testing of CCM+ bar and forgings, cast CoCr and low C CoCr bar, in the mock-sintered+solution-treated condition, are given in Table 7. The results indicate that the strength of CCM+ bar-stock and forgings are similar to each other and greater than that of cast CoCr, and the ductility of all three materials is similar. The tensile properties of low C CoCr bar in the same condition are also given for comparison. Note that this treatment causes greater reduction in the tensile and yield strength of low C CoCr compared to CCM+, so much so that the yield strength of low C CoCr in this condition is lower than that of cast CoCr in the same condition.

Table 6 - *Tensile properties of CCM+ forgings*

Condition	UTS, MPa	YS, MPa	Δl, %	RA, %
As-forged	1513±53	1175±85	10±4	14±4
Annealed (1093 °C/30 min)	1386	883	21	23
ASTM F 799 specification	1172	827	12	12

Table 7 - *Tensile properties of CoCr bar and forgings and cast CoCr in the mock-sintered+solution-treated condition*

Material	No. of Tests	UTS, MPa	YS, MPa	Δl, %	RA, %
CCM+ forging[1]	54	1030±66	596±56	18±3	15±3
CCM+ bar[1]	5	1132±60	650±36	17±3	18±2
Cast CoCr[2]	11	735±84	489±23	11±3	11±3
Low C CoCr bar[1,3]	4	815±9	409±12	33±2	30±2

[1] Mock-sintered+solution-treated, [2] Mock-sintered+HIP'ed+solution-treated
[3] From previous in-house testing

Porous-Coating Characterization

The average pore size of the porous-coating sintered on 12 CCM+ coupons by Thortex and AstroMet (6 by each supplier) was 272 ± 25 µm. The volume percent porosity of the coating on the 12 coupons was 58 ± 10%. The average thickness of the porous-coating was 0.96 ± 0.08 mm. The tensile pull-off strength of the porous-coating for 20 coupons sintered by Thortex and Astro Met was 60 ± 12 MPa. The static lap shear strength of the porous-coating for 12 coupons sintered by Thortex and Astro Met was 33 ± 5 MPa. The average lap shear strength and tensile pull-off strength were 65% and 200% respectively above the FDA guidance document [*13*] requirement of 20 MPa.

Fatigue Properties

Fatigue testing results are summarized in Table 8 and the corresponding logarithmic S/N curves are shown in Figures 9 and 10. The fatigue endurance limits, obtained using the 5 run-out criterion described in the Test Methods section, are given in Table 8 for sintered bead porous-coated CCM+ specimens from both vendors, smooth mock-sintered CCM+ specimens from both vendors, as well as comparison testing on cast (F 75) CoCr in the same two conditions.

Table 8 - *Ten million cycle fatigue endurance limits of CCM+ and cast CoCr, obtained using the 5 run-out criterion, from axial fatigue testing per ASTM E 466*

Condition	CCM+, MPa (sintering vendor A)	CCM+, MPa (sintering vendor B)	Cast CoCr, MPa (sintering vendor A)
Porous-coated	241	241	207
Mock-sintered, smooth	690	448	345

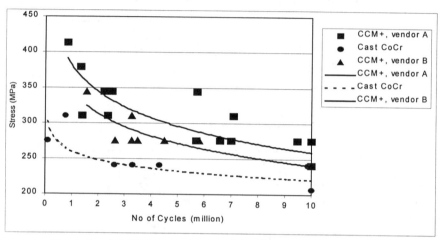

Figure 9 – *Axial fatigue testing results in the porous-coated condition*

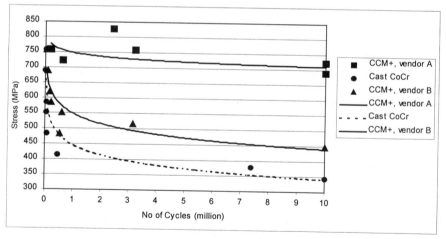

Figure 10 – *Axial fatigue testing results in the mock-sintered condition*

As shown in Table 8, the CCM+ specimens, porous-coated by either vendor, exhibited the same endurance limit of 241 MPa, based on the 5 run-out criterion. By comparison, the porous-coated cast CoCr specimens had an endurance limit of 207 MPa. In the smooth, mock-sintered condition the CCM+ specimens processed by one sintering vendor had an endurance limit of 690 MPa, and the specimens processed by the other sintering vendor had an endurance limit of 448 MPa. The smooth, mock-sintered cast CoCr specimens also had a lower endurance limit of 345 MPa. Thus, mock-sintered CCM+ smooth fatigue specimens show a fatigue strength improvement of at least 30% over the mock-sintered, smooth cast CoCr specimens. Porous-coated CCM+ exhibits a 16% improvement in fatigue strength over porous-coated cast CoCr.

Conclusions

1. The chemistry of CCM+ meets the requirements of the proposed new grade (grade 2, high C) of ASTM F 1537 wrought CoCr. The typical C content of this material is similar to that of cast CoCr.
2. CCM+ bar-stock was successfully forged into hip-stems. The forgings, after subsequent annealing, met the tensile property requirements of ASTM F 799.
3. Both CCM+ bar and forgings were successfully porous-coated. The average grain size of forged+porous-coated+solution-treated CCM+ hips ranges from 63 ± 19 μm to 145 ± 12 μm. By comparison, the average grain size of two commercially available porous-coated cast CoCr hip-stems from two different manufacturers was 2540 ± 395 and 3394 ± 876 μm respectively.
4. Metallography revealed that the cast CoCr implants contain the eutectic phase in the interdendritic regions. Literature [10,11] indicates that this phase consists of fcc Co, the intermetallic σ phase, and $M_{23}C_6$ and M_7C_3 carbides.
5. In the mock-sintered condition, the tensile and yield strength of CCM+ bar and forgings were similar to each other and greater than that of cast CoCr, and all three materials had similar ductility.
6. The average pore size of the porous-coating sintered on CCM+ coupons was 272 ± 25 μm. The volume percent porosity of the coating was $58 \pm 10\%$. The average thickness of the porous-coating was 0.96 ± 0.08 mm. The tensile pull-off strength of the porous-coating was 60 ± 12 MPa. The static lap shear strength of the porous-coating was 33 ± 5 MPa. The average tensile pull-off strength and lap shear strength were above the FDA guidance document [13] requirement of 20 MPa.
7. The 10 million cycle fatigue strength was 207 MPa for cast CoCr and 241 MPa for CCM+ bar in the porous-coated condition, and 345 MPa for cast CoCr and at least 448 MPa for CCM+ bar in the mock-sintered condition. Fatigue testing was performed only on bar since the tensile properties and microstructure of bar and forgings were similar, both in the porous-coated and the mock-sintered condition. Thus, the new grade of wrought CoCr exhibited higher fatigue strength than cast CoCr in both porous-coated and mock-sintered conditions. This may be attributed to the finer grain size of this material.

References

[1]   ASM handbook, ASM International, Materials Park, OH, 1993, vol 2, pp. 446.

[2]   Cobalt Facts, The Cobalt Development Institute, Wickford, Essex, U.K., August 1996.

[3]   Klarstrom, D. L., "Wrought Cobalt-Base Superalloys," *Journal of Materials Engineering and Performance*, 1993, vol. 2(4), pp. 523-530.

[4]   Sims, C. T., "A Contemporary View of Cobalt-Base Alloys," *Journal of Metals*, 1969, vol. 21(12), pp. 27.

[5]   Carpenter data sheet, Carpenter Technology Corporation, Reading, PA (2-95/4M)

[6]   Moral, F. R., "The Metallurgy of Cobalt Alloys - A 1968 Review," *Journal of Metals*, 1968, vol. 20, pp 52-59.

[7]   Wlodek, S. T., "Embrittlement of a Co-Cr-W (L-605) Alloy," *Transactions of the American Society for Materials*, 1963, vol. 56, pp. 287-303.

[8]   Clemow A. J. T. and Daniell, B. L., "Solution Treatment Behavior of Co-Cr-Mo Alloy," *Journal of Biomedical Materials Research*, 1979, vol. 13, pp. 265-279.

[9]   Kumar, P., Hickl, A. J., Asphahani, A. I., and Lawley, A., "Properties and Characteristics of Cast, Wrought, and Powder Metallurgy Processed Cobalt-Chromium-Molybdenum Implant Materials," in *Corrosion and Degradation of Implant Materials, ASTM STP 859*, A. C. Fraker and C. D. Griffin, Eds., American Society for Testing and Materials, West Conshohocken, PA, 1985, pp. 30-56.

[10] Kilner, T., Pilliar, R. M., Weatherly, G. C., and Allibert, C., "Phase Identification and Incipient Melting in a Cast Co-Cr Surgical Implant Alloy," *Journal of Biomedical Materials Research*, 1982, vol. 16, pp. 63-79.

[11] Kilner, T., Laanemae, W. M., Pilliar, R., Weatherly, G. C., and MacEwen, S. R., "Static Mechanical Properties of Cast and Sinter-Annealed Cobalt-Chromium Surgical Implants," *Journal of Materials Science*, 1986, vol. 21, pp. 1349-1356.

[12] Jacobs, J. J., Latanision, R. M., Rose, R. M., and Veeck, S. J., "The Effect of Porous Coating Processing on the Corrosion Behavior of Cast Co-Cr-Mo Surgical Implant Alloys," *Journal of Orthopaedic Research*, 1990, vol. 8, pp. 874-882.

[13] "Guidance Document for Testing Orthopaedic Implants with Modified Metallic Surfaces Apposing Bone or Bone Cement," U.S. Food and Drug Administration, April 28, 1994.

Kathy K. Wang,[1] Richard M. Berlin,[1] and Larry J. Gustavson[1]

A Dispersion Strengthened Co-Cr-Mo Alloy for Medical Implants

---

Reference: Wang, K. K., Berlin, R. M., and Gustavson, L. J., **"A Dispersion Strengthened Co-Cr-Mo Alloy for Medical Implants,"** *Cobalt-Base Alloys for Biomedical Applications, ASTM STP 1365,* J. A. Disegi, R. L. Kennedy, and R. Pilliar, Eds., American Society for Testing and Materials, West Conshohocken, PA, 1999.

Abstract: A dispersion strengthened version of forged Co-Cr-Mo alloy, GADS Vitallium® alloy, was developed to improve fatigue strength after sintering (1300°C), a required thermal cycle treatment for porous coated products. GADS alloy was made using a powder metallurgy process to create fine dispersed oxides in the alloy. The dispersed fine particles strengthen the alloy and prevent grain growth so that the alloy retains fatigue strength almost twice that of the standard forged alloy (ASTM F-799, Standard Specification for Cobalt-28 Chromium-6 Molybdenum Alloy Forgings for Surgical Implants) following porous coating sintering. The processing development, grain structures, fatigue properties, and corrosion resistance of the gas-atomized dispersion strengthened (GADS) alloy are reported here.

Keywords: Co-Cr-Mo alloys, dispersion strengthened, fatigue strength, porous coating, surgical implants

Vitallium® alloy is a cobalt-based alloy with chromium and molybdenum as major alloying elements. Both cast and forged Co-Cr-Mo alloys have proven to be extremely versatile implant materials. With over sixty years of clinical use, the cast alloy (ASTM F-75, Standard Specification for Cast Cobalt-Chromium-Molybdenum Alloy for Surgical Implant Applications) has established an exceptional record of biocompatibility and performance. The forged alloy (ASTM F-799, Standard Specification for Cobalt-28 Chromium-6 Molybdenum Alloy Forgings for Surgical Implants) exhibits unsurpassed strength and durability of fatigue resistance for total joint applications. However, with the introduction of porous coating to enhance non-cemented implant fixation, there was a concern over the fatigue strength of a forged

---

[1]Assistant Director, Principal Research Scientist, Director, respectively, Research and Development, Howmedica Inc., Pfizer Medical Technology Group, Rutherford, NJ 07070.

F-799 alloy substrate being weakened by porous coating sintering. The forged alloys obtain their high strength through deformation of their crystal structure by thermal mechanical processing such as rolling or forging. These strengthening effects are retained in the alloy at room temperature following forging. However, sintering a porous coating onto a conventional forged implant releases all the deformation energy and results in significant grain growth in the alloy microstructure. Consequently, the strength of the forged alloy drops significantly. This loss of strength precludes the use of forged F-799 alloy for porous coated products. The cast F-75 alloy has a much lower strength than the standard forged F-799 alloy. However, the strength of the cast alloy hot isostatically pressed (HIP'ed) after sintering is higher than that of forged F-799 alloy after porous coating sintering. For this reason, the cast F-75 alloy is still being used for many porous coated products.

It has been known for sometime that the strength of metals at high temperature could be increased by the addition of fine dispersed insoluble refractory oxides. One method that can produce such an alloy is the mechanical alloying method [1,2]. Studies [3] have shown that although a mechanically alloyed Co-Cr-Mo alloy had excellent fatigue strength (656 MPa at 10^7 cycles) after sintering, but there was insufficient ductility for conventional hot working.

A new dispersion strengthened Co-Cr-Mo alloy, GADS alloy [4-7], was developed by gas atomization to obtain a high post sintered fatigue strength of the oxide dispersion strengthened Co-Cr-Mo alloy [3] while maintaining the excellent hot workability of the forged alloy.

Materials

Processing of GADS Alloy

The GADS alloy is produced by powder metallurgy technique. The GADS powder has been produced using a gas atomizer. The raw materials used for making the forged Co-Cr-Mo alloy (ASTM F-799), were first melted in a vacuum induction furnace, followed by the introduction of a small amount of oxide forming elements, lanthanum (La) and aluminum (Al). Atomization was conducted immediately after the La and Al raw materials were melted into the alloy mixture. The molten metal was then atomized by nitrogen gas. The critical step during the atomization process was the final additions of La and Al. Once La and Al were melted into the bath, they quickly combined with oxygen and form oxides [6,7].

The GADS alloy powder was screened to remove the coarse +60 mesh (U.S. standard) powder. Typical GADS powder is shown in Figure 1. The −60 mesh powders were packed into mild steel can. Hot isotatic pressing (HIP) or extrusion was used to consolidate the powder. Conventional thermal mechanical processing (forging/rolling) was then applied to produce the GADS alloy. The processing parameters for the GADS alloy were reported previously [4,5]. The GADS alloy and its processing were patented for prosthesis applications [6,7].

Though the forging temperature used for the GADS alloy is normally 38°C higher than that for the forged F-799 alloy, the hot workability of GADS alloy was found to be excellent and comparable to that of forged Co-Cr-Mo alloy. The GADS alloy has been forged into hip prostheses at 1065-1230°C without any difficulties. Since the fatigue strength is much more critical than the tensile strength of an alloy for hip implant applications, the fatigue strengths of the GADS alloy in different conditions were determined.

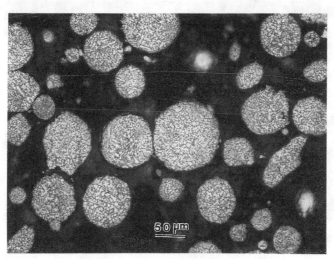

Fig. 1—*Microstructure of as-atomized GADS powder.*

Microstructure of GADS Alloy

Typical microstructures of as-forged and sintered GADS alloy are shown in Figures 2 and 3 respectively. The microstructure of as-forged GADS alloy is similar to that of forged F-799 alloy. Both the GADS alloy and the forged F-799 alloy exhibit a very fine equiaxed grain structure, grain size ASTM #10 and #9 respectively in accordance with ASTM E112, Standard Test Method for Determining Average Grain Size (Figure 2). GADS alloy tended to have finer grain size than the conventional forged alloy when forged at the same temperature. The grain size of GADS alloy remains fine (ASTM #8 or finer) even at a forging temperature of 1230°C. The forged alloy retains a fine grain structure and obtains its high strength through heavy deformation at a temperature below its recrystallization temperature.

After sintering, there is a significant difference in grain structure between the GADS alloy and the forged F-799 alloy (Figure 3). The GADS alloy still exhibited a fine, equiaxed grain structure (ASTM #6-8). However, a significant grain growth (ASTM #1) occurred in the forged F-799 alloy.

In addition to metallograpy, transmission electron microscopy (TEM) was also undertaken to characterize the nature of the dispersion strengthening mechanism of the sintered GADS alloy [4]. For comparison purposes, TEM work was also conducted on conventional ingot forged Co-Cr-Mo alloy as well as PM produced Co-Cr-Mo without the added oxide forming elements such as Al and La. The TEM study revealed that the

GADS alloy contained uniformly dispersed particles. Most of these particles ranged in size from 0.1 μm to 0.5 μm [4]. STEM/EDS analysis indicated that the dispersed particles were aluminum/lanthanum oxides. No dispersed particles were noted in the conventional ingot or P/M version of forged Co-Cr-Mo alloy [4].

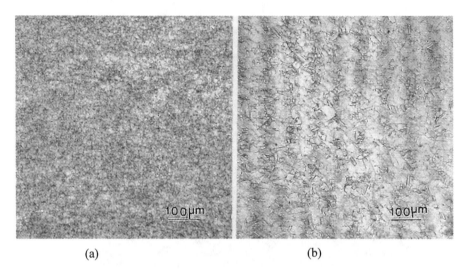

(a)   (b)

Fig. 2 – (a) Microstructure of as-forged GADS alloy (1120°C). (b) Microstructure of forged Co-Cr-Mo alloy in the as-forged condition (1120°C).

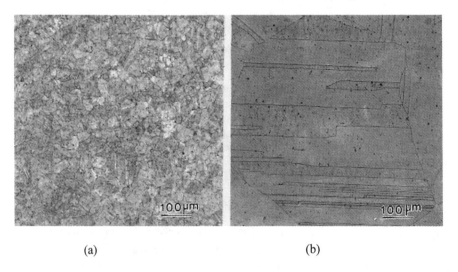

(a)   (b)

Fig. 3 – (a) Microstructure of sintered GADS alloy (1300°C). (b) Microstructure of forged Co-Cr-Mo alloy after sintering (1300°C).

Test Methods

Rotating Beam Fatigue Testing

The GADS alloy bars were subjected to smooth Krouse fatigue tests on a Fatigue Dynamics RBF 200 machine. The test mode was high cycle rotating beam fatigue. The test loading was constant force sinusoidal at approximately 167 Hz. The test stress ratio (R) was −1, i.e. fully reversed. All testing was conducted at room temperature in air. Test normally stopped when 10 million cycles were achieved. The fatigue strength is the stress to which the alloy can be subjected for 10 million cycles without failure. All Krouse fatigue specimens were machined from the as-forged (1120°C) or sintered (1300°C) bar stock. At least six specimens were tested for each condition. For comparison purposes, specimens made from the forged and the cast Co-Cr-Mo alloys were also tested under the same test condition as the GADS specimens. The purpose of this testing was to determine the effects of 1300°C sintering on the fatigue strength of GADS, forged and cast Co-Cr-Mo alloys.

Four-Point Bend Fatigue Testing

Four-point bend fatigue testing was performed on a Sonntag SF-01U test machine at a frequency of 30 Hz and a R value of 0.1. Testing was performed according to ASTM F-1160, Standard Specification for Constant Stress Amplitude Fatigue Testing of Porous Metal-Coated Metallic Materials. Fatigue strength was defined as the stress to which the alloy can be subjected for 10 million cycles without failure.

The porous coating is comprised of two layers of -20/+30 mesh (U.S. standard seive) spherical Co-Cr-Mo alloy powder. The porous coating powder was produced using cast F-75 alloy bar stock as conversion stock by the rotating electrode process (REP). The spherical F-75 alloy powder was applied in the green state using a non-toxic organic binder and sintered at 1300°C in vacuum.

At least six four-point bend specimens were tested with and without porous coating. The test specimens had a flat 50.8mm x 15.5mm gauge section (Figure 4). The calculation of fatigue strength was based on the substrate dimension of the test specimen without porous coating.

For comparison purposes, the cast alloy was also tested along with the GADS alloy under the same test conditions. This fatigue testing was conducted to determine the notch effects of porous coating on the fatigue strength of GADS and cast alloys.

Corrosion Testing

Corrosion testing was performed on the sintered GADS alloy. Specimens were in the form of 16 mm diameter x 3 mm thick disks. Tests were run in deaerated saline (0.9 wt.% NaCl) at 37°C. Each disk specimen was finished to a 600-grit silicon carbide immediately before testing.

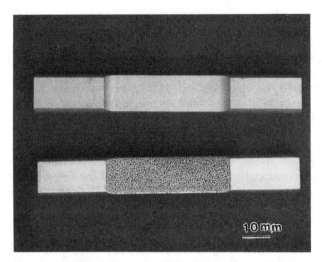

Fig. 4 – *A four-point bend test specimen with and without porous coating.*

ASTM G5, Standard Reference test Method for Making Potentiostatic and Potentiodynamic Anodic Polarization measurements was followed for the testing. The cast and forged Co-Cr-Mo alloys were also tested under the same condition. Four runs were conducted on the GADS alloy and three runs were conducted on the cast and forged alloys.

Results and Discussion

Rotating Beam Fatigue Testing

Smooth rotating beam fatigue strength for the GADS alloy compared with those of the cast and forged Co-Cr-Mo alloys is shown in Table 1. All alloys were tested in the condition before and after sintering. As can be seen, the cast alloy exhibits an initial 40% loss in fatigue strength after sintering and regains 85% of its original strength when hot isostatic pressing (HIP'ing, 1218°C at a pressure of 104MPa) and an additional solution anneal were employed following sintering. HIP'ing and solution annealing (1218°C with air cool or equivalent) after sintering are required for the cast alloy because they increase the ductility while improving the microstructure through closing of internal voids caused by the high temperature exposure required for sintering.

The forged Co-Cr-Mo alloy has a fatigue strength of 800 MPa, almost twice that of the cast alloy. However, after porous coating thermal cycles, a drop of close to 60% in fatigue strength was observed. Its post sintered fatigue strength (325 MPa) became lower than that (360 MPa) of the sintered and HIP'ed cast Co-Cr-Mo alloy. This dramatic loss of fatigue strength following sintering precludes the use of forged Co-Cr-Mo alloy for porous coated application. The strength of as-forged GADS alloy is superior to that of forged Co-Cr-Mo alloy when forged at the same temperature.

The GADS alloy retains a fatigue strength (620 MPa) almost twice that of the conventional forged alloy, following sintering, and 64% more than that of the cast Co-Cr-Mo alloy. This is because the dispersed oxides in GADS alloy are thermally stable and act as obstacles to inhibit the grain growth during the high temperature sintering [4-7].

Table 1 – *Smooth Rotating Beam Fatigue Strength of Cast, Forged and GADS Alloy*

Alloy	Condition	Fatigue strength (MPa)
Cast Co-Cr-Mo Alloy	Solution Treated (ST, 1218°C)	455
	ST (1218°C)+ Sintered (1300°C)	241
	Sintered +HIP (1218°C) +ST	360
Forged Co-Cr-Mo Alloy	As-forged (1066°C)	800
	Sintered (1300°C)	325
GADS Alloy	As- forged (1066-1232°C)	690-895
	Sintered (1300°C)	620

Four-Point Bend Fatigue Testing

The effect of porous coating on the fatigue strength of GADS alloy is shown in Table 2. For comparison purposes, the fatigue results for the cast Co-Cr-Mo alloy are also listed in Table 2. As can be seen, the smooth strength of GADS alloy is much stronger than that of the cast alloy after sintering. This corresponds to 76% improvement over the cast Co-Cr-Mo alloy after sintering. Porous coating does affect the strength of GADS alloy and the cast Co-Cr-Mo alloy. The GADS alloy is about 64% stronger than the cast Co-Cr-Mo alloy in the porous coated condition.

Corrosion Testing

The potential vs. normalized current data for the GADS alloy along with the cast and forged Co-Cr-Mo alloys are presented in Figure 5. The average scan of GADS alloy is almost identical with those of the cast and forged Co-Cr-Mo alloys. This indicates that the corrosion resistance of GADS alloy is comparable to that of cast or forged Co-Cr-Mo alloy.

Table 2 – *Fatigue Data of Cast and GADS Co-Cr-Mo Alloy*
(smooth sintered vs. porous coated)

Alloy	Condition	Fatigue Strength (MPa)
Cast Co-Cr-Mo Alloy	Smooth sintered[1]	380
Cast Co-Cr-Mo Alloy	Porous Coated	210
GADS Alloy	Smooth Sintered[2]	670
GADS Alloy	Porous Coated	345

[1] - Specimens were sintered +HIP +ST
[2] - Specimen went through sintering only

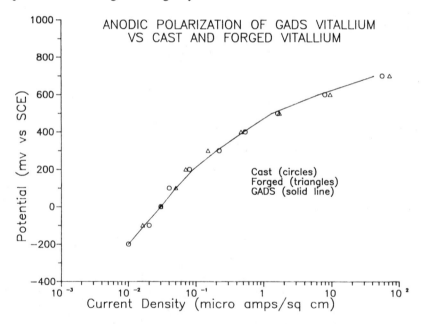

Fig. 5 — *Anodic polarization of GADS alloy, cast Co-Cr-Mo alloy and forged Co-Cr-Mo alloy.*

Summary

A dispersion strengthened version of forged Co-Cr-Mo alloy was developed for surgical implant applications. It yields a product with mechanical properties not attainable by conventional ingot metallurgy. It also exhibits an excellent corrosion resistance, comparable to the standard forged Co-Cr-Mo alloy.

References

[1] Benjamin, J. S., "Dispersion Strengthened Superalloys by Mechanically Alloying," *Met Trans.,* Vol.1, 1970, pp. 2943-2951.

[2] Weber, J. H., "High Temperature Oxide Dispersion Strengthened Alloys," *The 1980's- Payoff Decade for Advanced Materials*, Vol. 25, Science of Advanced Materials & Process Engineering Series, May 6-8, 1980, pp. 752.

[3] Andrews, H. L. and Gardiner, G. E., U.S. Patent 4,631,082, 20 Feb. 1985.

[4] Wang, K. K., Gustavson, L. J., and Dumbleton, J. H., "The Development of a New Dispersion Strengthened Vitallium® Alloy for Medical Implants," *Modern Developments in Powder Metallurgy*, Vol. 20, 1988, pp. 361-375.

[5] Wang, K. K., Gustavson, L. J., and Dumbleton, J. H., "A Dispersion Strengthened Vitallium® Alloy Developed for Medical Implants," *The 15th Annual Meeting of the Society for Biomaterials*, April 28-May2, 1989, pp. 154.

[6] Wang, K. K., Gustavson, L. J., and Dumbleton, J. H., U.S. Patent 4,668,290, 26 May 1987.

[7] Wang, K. K., Gustavson, L. J., and Dumbleton, J. H., U.S. Patent 4,714,468, 22 Dec. 1987.

H. E. Lippard[1] and R. L. Kennedy[1]

Process Metallurgy of Wrought CoCrMo Alloy

Reference: Lippard, H. E. and Kennedy, R. L., **"Process Metallurgy of Wrought CoCrMo Alloy,"** *Cobalt-Base Alloys for Biomedical Applications, ASTM STP 1365*, J. A. Disegi, R. L. Kennedy, and R. Pilliar, Eds., American Society for Testing and Materials, West Conshohocken, PA, 1999.

Abstract: This paper will describe the processing involved in the manufacture of low and high carbon grades of ASTM Wrought Cobalt-28 Chromium-6 Molybdenum Alloy for Surgical Implants (F 1537) as bar stock for use in a wide variety of biomedical applications. The effect of rolling temperature, annealing temperature, and cold drawing on the mechanical properties and grain size will be explored. Careful control of the thermo-mechanical working conditions is required to meet high strength requirements because age hardening and cold working of the alloy significantly degrade the ductility while producing the higher strength levels.

Keywords: CoCrMo, ASTM F 1537, mechanical properties

Introduction

The origins of today's CoCrMo alloy, as defined by ASTM F 1537, can be traced to the early 1930s while the first wrought version was introduced in 1952 [*1*]. Clearly, these dates preceded the introduction of vacuum melting. To meet the high standards required for biomedical applications, wrought alloy production follows a path very similar to that used for critical jet engine superalloys. Vacuum induction melting (VIM) is used as the primary melting stage provides high purity and excellent chemistry control. Secondary melting to improve alloy cleanliness and ingot structure is accomplished by the electroslag method. Modest ingot sizes of 13"-17" diameter are produced because of the relatively small section size of the final products and to minimize segregation problems. Ingots are typically homogenized, forged to an intermediate reroll billet, and then rolled to size. The material is supplied in the warm rolled or annealed condition depending on customer requirements. Straightening and peeling and/or centerless grinding of the bars to final size follow the rolling operation.

Mechanical properties of the alloy are strongly dependent on the microstructure and the working conditions. Final product typically has a uniform, fine grain size with the presence of small carbides and occasionally small amounts of sigma phase.

---

[1]Senior Engineer and VP of Technology, respectively, Allvac, 2020 Ashcraft Avenue, Monroe, NC 28111.

Melting

The primary melting process employed for cobalt-base alloy systems is vacuum induction melting (VIM). Raw materials for VIM melting must be carefully selected because of the limited refining capability of the VIM compared to the electric arc furnace-AOD. The charge make-up for CoCrMo alloy usually consists of virgin (elemental) raw materials, internal revert and purchased scrap. Use of purchased scrap is somewhat limited due to high carbon and phosphorus levels in cast CoCrMo. The VIM process melts material in a ceramic lined vessel at vacuum levels below 50 microns. Refining is by means of a carbon-oxygen boil, evaporation, desorption and in some cases metal/slag reactions. Undesirable elements such as oxygen and sulfur are typically held to very low levels (e.g. $\leq$ 30 ppm). Following the initial melt and refining periods, late additions are added and an in-process chemistry sample is taken and analyzed. Small additions of virgin raw materials can be used to adjust the composition range before casting. The CoCrMo alloy typically contains a purposeful nitrogen addition accomplished by means of nitrogen bearing manganese or chromium raw materials. While the VIM process does have limitations in terms of refining capability and restricts raw material usage, it has the great advantage of producing extremely tight chemistry distributions and very high yields. Typical heat sizes can range from 3,000 to 32,000 pounds.

Electroslag remelting (ESR) is the typical secondary melt process employed in cobalt-base alloys to further improve cleanliness, chemical homogeneity, and structure of the alloy. ESR is a consumable melt process where an electrode is supplied into a molten slag at a constant rate. The tip of the electrode melts and molten droplets are refined as they fall through the slag and then solidify in a water-cooled mold. A variety of slag compositions, typically in the CaF_2-CaO-Al_2O_3 family, is used depending on the alloy composition and desired refinement of the alloy. ESR is particularly effective for refining the oxide inclusions remaining after the VIM primary melt. In addition, desulfurization can occur in ESR.

Vacuum arc remelting (VAR) is also employed on occasion for selected alloy and application combinations. VAR refines the alloy composition by vacuum melting with a direct current arc into a water-cooled copper mold. Dissolved gases such as hydrogen and nitrogen are removed as well as high vapor pressure elements like the tramp elements lead, tin, and bismuth. The stability of many oxides prevents their direct removal in a gaseous form but flotation to the top of the resolidifying melt pool removes a significant fraction. VAR ingots exhibit smaller melt pools and less segregation during solidification due to the faster cooling rates when compared to ESR.

The typical CoCrMo ingot sizes produced by ESR and VAR are 13 to 17 inches in diameter and 3000 to 6000 pounds. The ingot diameters are relatively modest when compared to other superalloys produced using the same methods. Larger ingot diameters are not required because the typical final product size is less than 1.5 inches in diameter, which allows sufficient microstructural refinement during hot working to meet mechanical property requirements. The smaller ingot diameters also enhance the chemical homogeneity by higher solidification rates that minimize segregation.

Thermomechanical Processing

Ingots are homogenized before conversion to reduce microsegregation, which is inherent during solidification. Hot working is started by press or rotary hammer forging at temperatures in the 1700°F-2175°F range to produce reroll billets in the 4-5 inch diameter size. The deformation from the hot forging recrystallizes the coarse as-cast grain structure to a relatively fine size in the reroll billet. The surface of the reroll billet is conditioned by peeling or grinding to prevent any defects such as cracks or seams from the forging process carrying over to the final product. The billet may also be examined for internal defects using ultrasonic inspection at this stage.

Rolling to the final size product has traditionally been performed on hand rolling mills, which are labor intensive and require several reheatings of the material during the rolling process. More recently, continuous rolling mills have been introduced where long reroll billets are rolled in one operation to the final size.

The new continuous mills have multiple in-line stands that are controlled by one operator to work in synchronization while the billet is in contact with several stands simultaneously. The operator can vary the temperature profile during rolling through input billet temperature, rolling speed, and in-line water cooling boxes. The temperature profile variation tends to vary more smoothly on the continuous mill where no intermediate reheating occurs relative to the hand mills. Adiabatic heating from the rolling deformation produces heat in the material that is counterbalanced by thermal losses to the atmosphere, water cooling boxes, and the rolls. The amount of adiabatic heating is a function of the roll speed (strain rate), percent reduction at each stand, and the alloy composition. A specific rolling practice is developed for each alloy and sometimes subdivided based on the final product diameter to produce the required mechanical properties.

The continuous mills convert reroll billet 4 to 5 inches in diameter and 10 to 30 feet long to straight length bar from 0.5 to 4 inches in diameter and coils from 0.22 to 1.0 inches in diameter. The cooling rates available off the mill range from slow air cooling to water quench for both the bar and coil products. The higher capital cost of the continuous mill is offset by the lower labor costs and higher productivity.

Finishing and Inspection

The operations employed to produce finished mill products consist of cutting to length, straightening, removal of surface material, and heat treatment. The bar products are rough cut with automatic shears on the continuous rolling mill. Exact length cutting is performed with large abrasive saws. Out-of-straightness resulting from the hot rolling operation is removed by two-roll rotary straightening. The straightening operations are usually conducted at room temperature but can be performed at hot or warm temperatures. Straightening is often necessary after heat treatment, which releases residual stresses that warp the bars.

Surface defects from rolling such as laps, cracks, or seams are removed by coil shaving, bar peeling or centerless grinding. The typical envelope to eliminate surface defects and achieve final dimensional tolerances requires removal of 0.020-0.070 inches from the diameter. The exact amount depends on the alloy, processing path and final product diameter.

A variety of inspection methods are used to ensure the quality of the final mill product. Coil or bar/rod products can be inspected by eddy-current techniques for surface cracks as small as 25 μm deep. Eddy-current inspection is limited in penetration depth and can not provide full body inspection. Fluorescent dye penetrant testing is another surface defect inspection technique that can be easily applied to bar and rod products. The most complete level of inspection is offered by immersion ultrasonic. The ultrasonic technique can be manipulated to maximize surface defect sensitivity or full body sensitivity to detect internal cracks. Immersion ultrasonic inspection is the most expensive of the three techniques due to the capital equipment costs and relatively long time required to inspect each bar or rod.

Heat Treatment, Structure and Properties

Wrought CoCrMo is produced to two nominal chemistries (Table 1) that are differentiated only by their carbon content. The high carbon composition, which is used mainly for metal-on-metal implant devices, has lower ductility that requires closer control during hot working operations to prevent cracking, but it is still produced by conventional ingot metallurgy. The powder metallurgy route is used by some manufactures to produce this grade.

Table 1 – *Nominal Compositions of High and Low Carbon CoCrMo Alloys*

Element, wt%	Low Carbon	High Carbon
Cobalt	Bal.	Bal.
Chromium	27.7	27.7
Molybdenum	5.7	5.7
Manganese	0.7	0.7
Iron	0.5	0.5
Nickel	0.7	0.7
Silicon	0.7	0.7
Nitrogen	0.18	0.18
Carbon	0.05	0.25

CoCrMo is a work strengthened alloy and cannot be suitably hardened by heat treatment alone. Careful control of the final rolling practice is required to achieve mechanical properties. (Figure 1) displays the mechanical properties obtained from early hand mill experiments that evaluated the effect of percent reduction and rolling temperature. The mechanical properties are insensitive to the percent reduction in the range investigated. A minor effect of rolling temperature in the 2000°F-2100°F range is

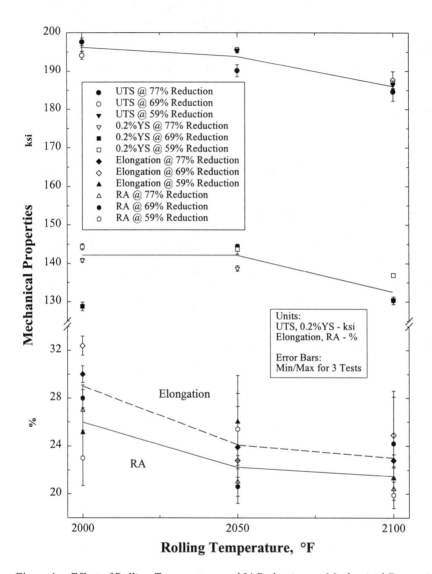

Figure 1 – *Effect of Rolling Temperature and % Reduction on Mechanical Properties of Low C CoCrMo Bar*

exhibited by a greater combination of strength and ductility at 2000°F than at the higher temperatures. The high carbon alloy produces similar behavior but is more sensitive to forming carbide films on grain boundaries at the higher rolling temperatures, where carbon is partially solutioned.

The controlling parameters for continuous mills are different from the hand mills and their specific values for each material are closely held by alloy producers. In general terms, the rolling speed, reduction and amount of water cooling determine the temperature profile of the material at each stand. These factors in turn control the microstructure and amount of stored strain energy that determine the mechanical properties. (Table 2) compares mechanical properties with 95% confidence intervals typical of one inch diameter round bar from a continuous rolling mill to the ASTM F 1537 requirements.

Table 2 – *Typical Mechanical Properties of ASTM F 1537 Alloy Bar*

Condition	Ultimate Tensile Strength, ksi	0.2% Yield Strength, ksi	Elongation, %	Reduction in Area, %
Low Carbon – Annealed	162.6 ±2.8	88.6 ±1.7	28.5 ±2.0	24.1 ±1.2
Low Carbon – Warm Worked	194.1 ±1.0	146.5 ±1.2	21.4 ±0.5	19.2 ±0.4
High Carbon – Warm Worked	193.5 ±1.0	146.7 ±1.3	15.2 ±0.8	13.7 ±0.9
ASTM F 1537 Specification				
Annealed	130	75	20	20
Hot Worked	145	101	12	12
Warm Worked	170	120	12	12

The mechanical properties from continuous mill processed bar exceed the ASTM minimum requirements by a margin that allows some small processing variations to occur without jeopardizing the material qualification. The high carbon version possesses the same strength level as the low carbon in a warm worked condition but has lower ductility due to more grain boundary carbide precipitates. The carbides are not effective as a strengthening agent in the warm worked condition where the grain size and stored strain energy due to deformation are the primary strengthening mechanisms. The grain size effect on strength is governed by the classic Hall-Petch relationship where higher strengths are produced by finer grain sizes. Stored strain energy imparted by finish rolling in the warm working temperature range is the most important strengthening mechanism. The working temperature range is restricted at a lower temperature bound by the tendency for crack initiation and at an upper bound by rapid dislocation motion. The effectiveness of the warm working on annealed material is summarized in (Table 3). The mechanical properties show the classic exchange of higher strength for lower ductility as the warm working temperature is decreased. Since no recrystallization and no change in grain size was apparent for the 1700°F and 1800°F samples, all of the strengthening can be attributed to stored strain energy.

Strengths greater than the warm worked condition can be reached by further increasing the stored strain energy through cold work. The cold worked mechanical properties shown in (Table 3) were obtained from 0.390 inch diameter rod that had been drawn with a 20% reduction from the as-rolled condition. Cold working increased strength levels by 70 ksi while decreasing elongation and RA by 10% and 1%, respectively.

Table 3 – *Effect of 20% Reduction on Mechanical Properties of Low C CoCrMo Rod*

	Cold Worked	Warm Worked @		
		1700°F	1800°F	1900°F
UTS, ksi	256.5 ±5.0	214.7 ±3.0	214.1 ±3.0	206.3 ±3.0
0.2%YS, ksi	208.4 ±5.0	170.0 ±3.0	159.9 ±3.0	148.8 ±3.0
Elongation, %	10.8 ±2.0	14.6 ±2.0	24.1 ±2.0	30.6 ±2.0
Reduction in Area, %	19.0 ±2.0	18.8 ±2.0	24.6 ±2.0	29.8 ±2.0
Hardness, HRC	48 ±1.0	48 ±1.0	46 ±1.0	45 ±1.0

Alternatively, an age hardening response by carbide precipitation can provide an increment to the strength levels [2]. The peak age strength increase is less than the increment produced by 20% cold work and the carbide precipitation reduces the elongation to 5%. The low ductility levels severely hamper manufacturability to final product form and limit the possible applications.

Strength levels lower than the warm worked condition are produced by annealing the alloy after deformation. The typical annealing temperature is in the 2000°F-2050°F range. Grain growth occurs at these annealing temperatures, which increases the grain size to the ASTM 5-6 range. The effect of annealing temperature on mechanical properties of warm worked material is shown in (Figure 2). Properties are generally found on one of two plateau regions. Warm worked property levels remain almost unchanged up to 1850°F then drop significantly at 1900°F and above. The yield strength shows a continuing decline after passing the critical annealing temperature but all other mechanical properties remain nearly constant.

Grain growth closely parallels the mechanical properties as shown in (Table 4). The warm worked ASTM 11 grain size is stable to 1800°F and then transitions in stages to the typical annealed grain size of ASTM 5 as the annealing temperature increases. A duplex grain structure is observed at 1850°F where some grains have grown to an ASTM 7 while others have coarsened only slightly to ASTM 10. The 50°F window between the two conditions is not useful in a commercial manner because most industrial furnaces are capable of ±25°F temperature control.

Table 4 – *Grain Size of 1 Hour Annealed Low C CoCrMo Alloy, 0.437 Inch Diameter Coil*

Anneal Temperature	1600°F	1700°F	1800°F	1850°F	1900°F	2000°F
ASTM Grain Size	11	11	11	10 / 7	8	5

The typical microstructure of wrought CoCrMo bar, as shown in (Figure 3), consists of a uniform, equiaxed fine grain size of ASTM 10 to 12. The matrix is a metastable fcc crystal structure with a distribution of very fine carbides. The chromium-rich $M_{23}C_6$ carbide is the most commonly observed, but molybdenum-rich M_6C and MC have also been reported [3]. Occasionally, large blocky particles, significantly bigger than the carbides are observed. (Figure 4) shows both of these features. SEM analyses of these large particles show them enriched in chromium, molybdenum and silicon compared to the matrix, suggesting they are sigma phase [4]. These particles are only partially solutioned with

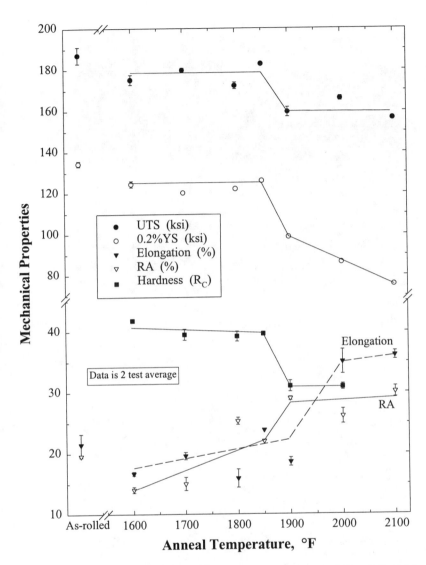

Figure 2 – *Effect of Annealing Temperature on Warm Worked Low C CoCrMo Alloy*

heat treatments as high as 2200°F-1 hour. The high carbon version of CoCrMo has a substantially larger number of carbides, but this is the only difference in structure for the two variants of the wrought alloy. The very small size of the carbides (1-≤3 microns) in wrought low carbon CoCrMo, suggest that much larger carbides that would have been present in the cast ingot, have been refined by solutioning and reprecipitating during thermomechanical working. The annealed structure is significantly coarser in grain size as shown in (Figure 5). Note the absence of carbides and the presence of many annealing twins and stacking faults in many of the grains.

Figure 3 – *Warm Worked Structure of Low C CoCrMo Bar*

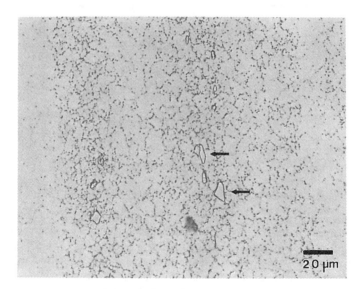

Figure 4 – *Warm Worked Structure of Low C CoCrMo Bar Revealing Carbide and Sigma Phase (arrows)*

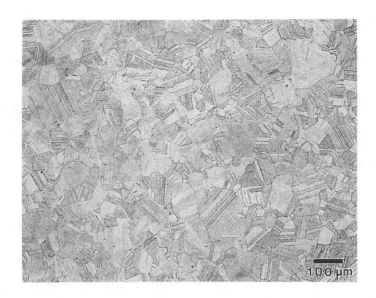

Figure 5 – *Annealed Structure of Low C CoCrMo Bar*

Conclusions

Wrought CoCrMo alloy has been in production for many years by the VIM-ESR method. This premium ingot metallurgy route is well suited to produce both the high and low carbon versions of the alloy. The processing employed results in a very consistent chemistry and extremely fine grained, uniform structure. Mill product in the form of bar, rod and coil has uniform properties that consistently exceed the most common industrial specification, ASTM F 1537.

References

[1] "Strength for Life: The Vitallium® Alloy Story," Howmedica Product literature, http://www.osteonics.com/howmedica/book/page01.html, 1999.

[2] Taylor, R. N. and Waterhouse, R. B., "A Study of the Ageing Behaviour of a Cobalt Based Implant Alloy," Journal of Materials Science, vol. 18, 1983, pp. 3265-3280.

[3] Immarigeon, J-P., Rajan, K., Wallace, W., "Microstructural Changes During Isothermal Forging of a Co-Cr-Mo Alloy," *Metallurgical Transactions A*, vol. 15A, 1984, pp. 339-345.

[4] Sims, C. T., "Cobalt-Base Alloys", *The Superalloys*, C. T. Sims and W. C. Hagel, Eds., John Wiley & Sons, New York, 1972, pp. 145-173.

Armando Salinas-Rodríguez[1]

The Role of the FCC-HCP Phase Transformation During the Plastic Deformation of Co-Cr-Mo-C Alloys for Biomedical Applications

Reference: Salinas-Rodríguez, A., **"The Role of the FCC-HCP Phase Transformation During the Plastic Deformation of Co-Cr-Mo-C Alloys for Biomedical Applications,"** *Cobalt-Base Alloys for Biomedical Applications, ASTM STP 1365*, J. A. Disegi, R. L. Kennedy, R. Pilliar, Eds., American Society for Testing and Materials, West Conshohocken, PA, 1999.

Abstract: The plastic deformation behavior of solution treated Co-27Cr-5Mo cast alloys with C contents up to 0.31 wt% were studied at 298 K by means of uniaxial tension and compression tests. In addition, the effect of initial grain size on the plastic flow behavior of annealed Co-27Cr-5Mo-0.05C wrought alloy was investigated. Optical and scanning electron microscopy and X-ray diffraction techniques were used to characterize the microstructures produced by plastic deformation.

Carbon contents greater than 0.05 wt% and furnace cooling after solution treatments at 1250 °C were found to inhibit the occurrence of strain-induced fcc(metastable)=>hcp phase transformation and lead to high strain hardening rate, high strength and low ductility. A similar behavior was found in annealed, 0.05 wt% C alloy for grain sizes smaller than 10 μm. In contrast, a fast cooling rate after the solution treatment and grain sizes in the range 100-2000 μm for this later alloy were found to promote phase transformation during deformation which lead to a rapid decrease of the rate of strain hardening. The yield strength and the strain hardening behavior of both cast-solution treated and wrought-annealed materials were found to depend on the type of loading during the tests.

The results allowed to conclude that the strain-induced fcc(metastable)=>hcp phase transformation plays an important role during large strain plastic deformation of low carbon Co-27Cr-5Mo implant alloys. However, when the carbon C in these alloys is larger than 0.05 wt%, the ductility and the fracture behavior are controlled by the size, morphology and distribution of secondary phase carbide particles in the microstructure.

Keywords: Plastic deformation, strain hardening, microstructure, cobalt-base alloys

[1] Professor, CINVESTAV-Unidad Saltillo, P.O.Box 663, Saltillo, Coahuila, México 25000

Introduction

Co-Cr-Mo-C alloys used in the manufacture of orthopedic prosthetic devices possess very low stacking fault energies. As a result, plastic deformation by dislocation glide in the face centered cubic (fcc) metastable phase is severely restricted [1-4]. This behavior leads to formation of strain-induced crystal defects, such as intrinsic stacking faults, twins and regions of highly localized slip along pre-existing and strain-induced stacking faults, when these alloys are subjected to external stresses exceeding their yield strength. Thus, the ductility of these materials, particularly in as-cast and solution treated conditions, is small compared with other fcc metals. The interactions between dislocations of limited mobility and dissociated dislocations and/or twins lead to very rapid and highly localized hardening. This eventually leads to fracture if no additional strain-producing mechanisms become available to relieve the high stresses required to maintain plastic flow. Large strain plastic deformation of Co-Cr-Mo-C alloys must therefore be microscopically accommodated by alternative mechanisms. Twinning and strain-induced phase transformations have been suggested [1-3] to play an important role during large strain plastic deformation in cast and solution treated Co-Cr-Mo-C alloys, although their contribution to ductility improvements is still unclear.

The phase transformation from the high temperature fcc phase to the low temperature hexagonal close packed phase (hcp) in Co-base alloys takes place via a martensitic mechanism. The kinetics of the transformation is, however, very sluggish and large amounts of the high temperature fcc phase remain in metastable form after cooling to room temperature. Recent work [5] has shown that the metastable fcc phase in homogenized and quenched Co-27Cr-5Mo-0.05C alloy can fully transform to hcp during six hours of isothermal aging at temperatures between 800 and 870 °C. Alternatively, the fcc(metastable)=>hcp phase transition can be induced dynamically during plastic deformation. Graham and Youngblood [6] demonstrated that the amount of hcp phase formed during cold swaging of Co-Cr-Mo-Ni multiphase alloys increased rapidly with the amount of deformation. The tensile yield strength was correlated linearly with the amount of strain-induced hcp phase. Olson and Cohen [7] have shown that the ability of low stacking fault energy materials to deform uniformly during tensile testing is enhanced by strain-induced phase transformations. For example, the ductility of TRIP (Transformation Induced Plasticity) steels increases considerably when martensite forms dynamically during plastic deformation. Previous work [8] on the deformation behavior of wrought Co-27Cr-5Mo-0.05C alloy showed that increasing the initial grain size of the metastable fcc phase from 7 to 70 μm improves the tensile uniform ductility by nearly 100%. This ductility enhancement was attributed to the occurrence of transformation-induced plasticity in the coarse grained material.

This article examines the role of the strain-induced fcc(metastable)/hcp phase transformation in the large strain plastic deformation of a wrought and annealed Co-27Cr-5Mo-0.05C alloy and three solution treated Co-27Cr-5Mo cast alloys with carbon contents in the range 0.05 to 0.31 wt%. The role of the alloy composition, in particular the effect of carbon content, in promoting or inhibiting the strain-induced martensitic transformation is not well known. Preliminary unpublished experimental results suggested that increasing the C content of the solution treated Co-Cr-Mo cast alloys inhibit the strain-induced transformation. This article attempts to clarify this issue.

Materials and Procedures

Experimental Alloys

The chemical compositions of the four experimental alloys are given in (Table 1). Materials designated C1, C2 and C3 were investment cast into 13.5 mm diameter by 250 mm long bars by remelting low and high carbon commercial ASTM-F75-type alloys in a vacuum induction furnace. Material W was a 35 mm diameter, press forged, implant quality Co-27Cr-5Mo-0.05C alloy round bar.

Table 1 *Chemical Analyses of the Experimental Alloys*

Material	Solute Content (wt%)							
	C	Cr	Mo	Ni	Fe	Mn	Si	Co
C1	0.05	28.10	5.50	0.17	0.16	0.77	0.51	Bal.
C2	0.16	30.21	4.90	0.37	0.19	0.67	0.50	Bal.
C3	0.31	30.45	4.23	0.45	0.20	0.60	0.70	Bal.
W	0.05	28.97	5.59	0.21	0.17	0.79	0.49	Bal.

Heat Treatments

The as-cast materials (C1, C2 and C3) were solution treated at 1250 °C during 2 hours in a tube chamber furnace under an atmosphere of high purity argon gas. After heat treating, some of the bars were quenched into water at room temperature and the remaining bars were allowed to cool inside the furnace to 300 °C before quenching. Small lots of material W were annealed in the same furnace at temperatures in the range 500-1250 °C. Annealing times were varied from a few minutes up to 24 hours to produce different grain sizes after cooling to room temperature at 10 K/min. Five different grain sizes were selected (annealing conditions are giving in parenthesis) to carry out the present study: 7.5 μm (500 °C, 2h), 9.5 μm (500 °C, 24 h), 100 μm (1100 °C, 2 hours), 878 μm (1250 °C, 2 hours) and 2156 μm (1250 °C, 24 hours).

Characterization of Plastic Flow Behavior

Constant true strain rate tension and compression tests were performed on an automated, computer-controlled, mechanical testing system at strain rates of 2.1×10^{-2} s^{-1} for material W and 1×10^{-2} s^{-1} for materials C1, C2 and C3. The test specimens were machined after the heat treatments to have their axes parallel to the longitudinal direction of the original bars. Tensile specimens had a gauge length of 25.4 mm and a diameter of 6.3 mm while compression specimens were 12 mm long and 8 mm in diameter. The applied force was measured using an 150 kN tension-compression load cell. The strain was measured directly on the tension specimens using a calibrated axial extensometer. In the case of compression the strain was estimated from measurements of the gap between the compression platens using a position sensor attached to the crosshead of the testing system. These measurements were also used as feedback signal to vary the crosshead velocity according to the change in dimensions of the deforming specimens to maintain a constant true strain rate. All tension and compression tests were performed on duplicate specimens.

Characterization of the Deformation Microstructure

The microstructures of the deformed materials were characterized by standard metallographic techniques using reflected light and scanning electron microscopy. Specimen preparation for metallography included a final etching step during 15 seconds in a 6:1 solution of HCl and H_2O_2. The structure of the materials was characterized by X-ray diffraction on samples prepared metallographicaly. 2θ scans between 40° and 55° were carried out using monochromatic Cu Kα radiation. The X-ray source was operated at 45 kV and 40 mA. The intensity ratio between Kα_1 and Kα_2 was 0.5. The X-ray patterns were recorded with the collimations of a 1° divergent slit and a 0.05 mm wide detector slit at a scan rate of 0.02 °2θ/s. The 2θ angular scan range selected allows to determine unambiguously the coexistence of the fcc and hcp cobalt phases; within this region the strong $(200)^{fcc}$ and the $(10\bar{1}1)^{hcp}$ diffraction peaks are well isolated and do not overlap with any other diffraction peak. The relative amounts of fcc and hcp phases in the deformed specimens were estimated from the integrated intensities of these diffraction peaks using the method of Sage and Gillaud [9].

Experimental Results and Discussion

Initial Microstructure

Wrought Alloy - The microstructure of material W after annealing consisted of a single phase aggregate of equiaxed grains containing annealing twins (Figure 1). The grain size increased with annealing temperature and time but the general features of the microstructure did not change. The X-ray diffraction patterns of the annealed specimens (Figure 2) showed no evidence that the fcc(metastable)=>hcp transformation occurred after the slow cooling from the annealing temperatures. However, recrystallization of the fcc phase produced sharper and more intense $(111)^{fcc}$ diffraction peaks at higher annealing temperatures. The significant broadening observed in the $(200)^{fcc}$ diffraction peak at the highest annealing temperature is due to the effect of the coarser fcc grain size produced by this heat treatment.

Solution Heat Treated Cast Alloys - In general, the microstructures of the cast alloys (C1, C2, C3) after the solution treatment consisted of a dispersion of second phase carbide particles in a matrix of very large recrystallized grains. The amount of second phase particles increased significantly with carbon content and decreasing cooling rate.

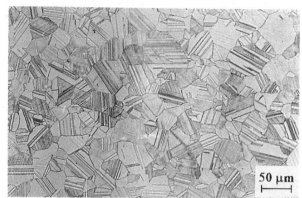

Figure 1 - *Microstructure of Co-27Cr-5Mo-0.05C alloy annealed 2 hours at 1100 °C.*

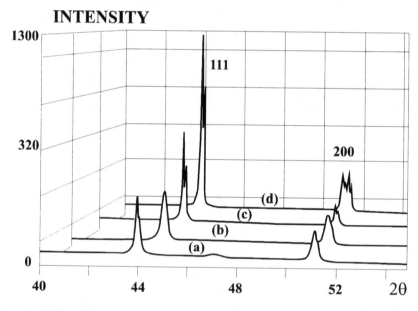

Figure 2 - *Effect of annealing temperature on the X-ray diffraction patterns of wrought Co-27Cr-5Mo-0.05C alloy (a) as-received, (b) 500 °C, (c) 1100 °C and (d) 1250 °C.*

The most significant effect of the cooling rate was observed on the morphology and distribution of the second phase particles in alloy C3 with 0.31 wt% C (Figure 3). In this material the grain boundary phase after slow cooling consisted of a lamellar pearlitic-type constituent. In addition, roughly squared particles were formed in the interior of the grains. In contrast, water quenching from 1250 °C (Figure 3b) produced a semi continuous grain boundary film and globular intraganular particles. Observation of this later microstructure at higher magnifications showed that the globular particles and the intergranular film developed a peculiar serrated interface with the surrounding matrix. Kilner et al. [*10*] have associated this type of interface to quenching effects on the incipiently melted eutectic present in Co-Cr-Mo high carbon alloys heat treated at temperatures above 1230 °C K.

The intragranular particles in both quenched and furnace cooled samples of material C3 varied in size from 1-20 μm. X-ray energy dispersive microanalysis (EDAX) of several particles showed that they are complex carbides containing Co, Cr, Mo and Si.

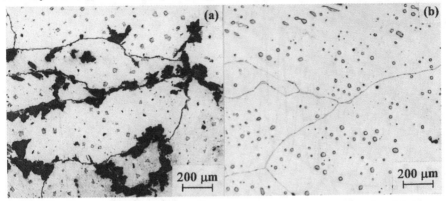

Figure 3 - *Microstructures of alloy C3 (0.31 wt% C) solution treated at 1250 °C (a) furnace cooled, (b) water quenched.*

The pearlitic constituent and the grain boundary serrated film were found to be slightly richer in Cr and Mo than the Co-matrix due to preferential segregation of these elements to the interdendritic regions during solidification and to the low diffusivity of Cr and Mo in the fcc solid solution matrix. The intragranular carbides were found to be poorer in Co but richer in Mo when compared with the carbides formed at the grain boundaries.

X-ray diffraction patterns of all three solution treated cast alloys were similar to those observed in the annealed wrought alloy (Figure 2) and only exhibited the presence of reflections corresponding to the fcc phase. This result indicates that water quenching or furnace cooling of alloys C1, C2 and C3 do not cause the fcc/hcp transformation and the high temperature fcc phase remains in metastable form after cooling to room temperature.

Deformed Microstructure

Wrought Alloy - The fractured tensile specimens were examined by x-ray diffraction and optical metallography on a cross-section near the fracture surface. The microstructure revealed that a significant amount of strain-induced hcp phase was formed during deformation. This effect is illustrated in (Figure 4) by the large increase in the intensity of the $(10\overline{1}1)^{hcp}$ diffraction peak and the decrease in the intensity of the $(200)^{fcc}$

diffraction peak.. The strain-induced fcc(metastable)=>hcp phase transformation is accompanied by a massive formation of intragranular striations within the prior fcc grains (Figure 5). The weight fraction of hcp phase, estimated from the diffracted intensities of the $(200)^{fcc}$ and $(10\overline{1}1)^{hcp}$ peaks in the patterns shown in (Figure 4), was found to remain constant for initial fcc grain sizes greater than 10 μm.

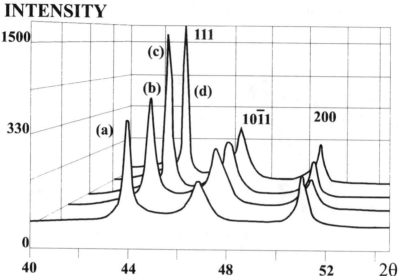

Figure 4 - *Effect of tensile deformation to fracture on the X-ray diffraction patterns of annealed Co-27Cr-5Mo-0.05C alloy with different grain sizes: (a) 7.5 μm, (b) 9.5 μm, (c) 100 μm and (d) 878 μm.*

Figure 5 - *Microstructure of Co-27Cr-5Mo-0.05C (initial fcc grain size: 100 μm) strained in tension to fracture.*

In the case of specimens deformed in compression, the evolution of the microstructure was followed as a function of strain. The effect of compressive strain on the weight fraction of strain-induced hcp phase as a function of grain size is shown in (Figure 6). As can be seen, after an incubation strain of about 0.05, the amount of strain-induced hcp phase increases linearly with strain. Although the experimental uncertainty of individual data points in (Figure 6) is large due to the effects of plastic deformation, grain size, etc., on the diffraction peaks, the data do suggest that a certain amount of plastic deformation must occur in the fcc parent phase before the hcp martensite can be formed.

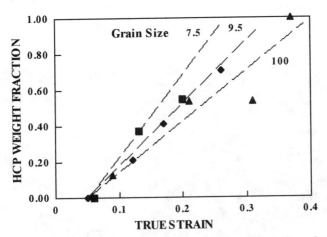

Figure 6 - *Effect of initial fcc grain size (given in μm) on the amount of strain-induced hcp phase formed during compressive deformation of annealed Co-27Cr-5Mo-0.05C.*

In addition, the data show that, for a given plastic strain, the amount of strain-induced hcp phase increases as the initial grain size of the fcc phase decreases. This later result suggests that plastic deformation by dislocation glide is easier in material with coarser initial fcc grain size. In contrast, in material with finer initial fcc grain size, the increased number of grain boundaries imposes an additional restriction to dislocation glide. The results presented up to now indicate that loading of Co-27Cr-5Mo-0.05C alloy causes the simultaneous activation of two different strain-producing mechanisms during plastic deformation: dislocation glide and the fcc-hcp phase transformation. Dislocation glide is inherently difficult in the present material due to its low stacking fault energy. However, the results show (Figure 6) that, as the initial fcc grain size increases, dislocation glide becomes relatively easier and, as a result, the amount of strain-induced phase transformation required to relieve the imposed stress is smaller than for fine grained material. This microstructure evolution has profound effects on the flow and strain hardening behaviors of this alloy.

Solution Heat Treated Cast Alloys - The dominant features of the deformation microstructures in the solution treated cast alloys was the development of numerous intersecting deformation bands and extensive second phase particle cracking (Figure 7). These features were a general result of the plastic deformation of the materials tested irrespective of carbon content and cooling rate after the heat treatment. Vander Sande et al. [1] observed similar deformation bands by transmission electron microscopy in Co-27Cr-5Mo-0.30C water quenched from 1230 °C. Rajan [2] characterized the deformation substructures produced by plastic tensile deformation up to strains of about 0.12 by transmission electron microscopy. Analysis of the substructures revealed that plastic deformation generated large densities of intrinsic stacking faults with the consequent fault overlapping and intersection as the strain was increased. At larger strains twinning in the fcc matrix, as opposed to strain-induced fcc/hcp phase transformation, was considered the relevant strain producing mechanism.

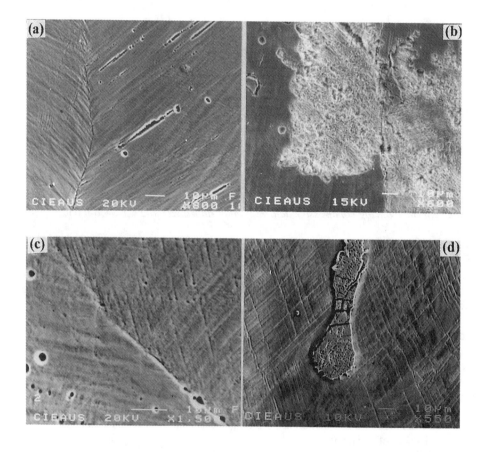

Figure 7 - *Deformation microstructures of solution treated cast alloys: (a) alloy C1 furnace cooled, (b) alloy C3 furnace cooled, (c) alloy C1 water quenched and (d) alloy C3 water quenched.*

Figure 8 shows the X-ray diffraction patterns obtained from the water quenched Co-27Cr-5Mo alloys with 0.05 and 0.31 wt% C deformed to compressive strains of about -0.2. It is noteworthy that the water quenched 0.05 wt% C alloy exhibits a significant increase in the intensities of the $(10\bar{1}0)^{hcp}$ and $(10\bar{1}1)^{hcp}$ diffraction peaks while, at the same time, the intensity of the $(200)^{fcc}$ diffraction peak decreased.

Considering that before deformation there was no hcp phase present in the microstructure, it becomes clear that the contribution of the fcc(metastable)=>hcp phase transformation to the plastic deformation of solution treated Co-27Cr-5Mo cast alloys is greater when their carbon content is small (close to 0.05 wt%). An estimation of the relative weight fractions of dynamically formed hcp phase gives 12% and 57% for the 0.31 and 0.05 wt% C alloys, respectively. This result indicates that the carbon content in Co-Cr-Mo alloys plays an important role on their ability to accommodate macroscopically imposed deformations via strain-induced phase transformation. Thus, significant improvements in ductility can be obtained in cast and solution treated alloys by decreasing their C content. However, the increase in ductility will result not only from the reduced amounts of hard and brittle second phase carbide particles. An additional contribution to the ductility of the alloy results because the material becomes capable of deforming plastically via strain-induced phase transformation.

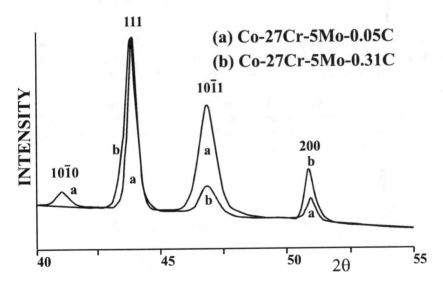

Figure 8 - *X-ray diffraction patterns obtained after compressive deformation (-0.2) of alloys C1 and C3 water quenched after solution treatments at 1250 °C.*

Effect of the Strain-induced fcc-hcp Phase Transformation on Strain Hardening

Wrought Alloy - The type of deformation and the initial fcc grain size affect both the yield strength and the strain hardening of annealed Co-27Cr-5Mo-0.05C alloy. The evolution with stress of the strain hardening rate ($\theta = \delta\sigma/\delta\varepsilon/E$, E=Young modulus) (Figure 9) shows a rapid decrease in θ from the proportional limit to the yield stress. At larger stresses, the strain hardening rates reach a steady state value which apparently does not depend on grain size and deformation mode. This steady state flow regime ends with fracture.

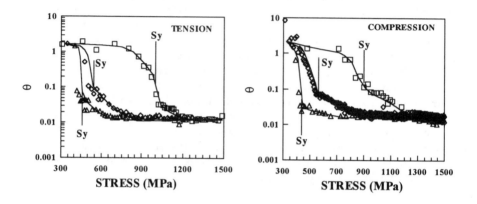

Figure 9 - *Effects of mode of deformation and initial fcc grain size (squares: 9.5 μm, diamonds: 100 μm, triangles: 878 μm) on the stress dependence of the strain hardening rate of annealed Co-27Cr-5Mo-0.05C alloy. Sy=Yield Strength.*

In the case of the tension tests performed, plastic deformation was uniform throughout the specimen gauge length until fracture occurred on a plane normal to the tensile axis. The weight fraction of strain-induced hcp phase and the fracture strains were independent of the initial fcc grain size for grain sizes greater than ~10 μm (Figure 10).

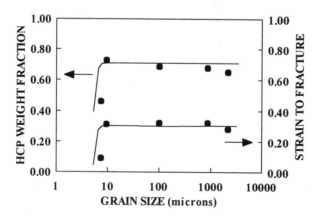

Figure 10 - *Relationship between strain to fracture and amount of strain-induced hcp phase formed during tensile deformation of annealed Co-27Cr-5Mo-0.05C alloy.*

The microstructure data presented earlier (Figures 4-6) showed that plastic deformation of Co-27Cr-5Mo-0.05C alloy causes the transformation of the metastable fcc phase to the hcp phase. Dislocation glide in this material is difficult due to the low stacking fault energy of Co-base alloys. As a result, rapid strain hardening can lead to premature fracture if no additional strain-producing mechanisms become available during plastic deformation. Therefore, the strain-induced martensitic fcc-hcp phase transformation plays a double role during plastic deformation in this material. First, the fcc-hcp transformation helps to relieve the rapid build-up of internal stresses caused by the relative inability of the fcc phase to deform plastically via dislocation glide. Second, the shape change associated with the transformation provides the material with an additional strain-producing mechanism and inhibits the formation of inernal cracks.

The microstructure evolution and strain hardening behavior observed (Figures 4-6, 9) indicate that the formation of the hcp phase during plastic deformation acts as a soft strain producing mechanism at low stresses and causes a rapid decrease in the strain hardening rate. However, as deformation proceeds, the flow stress increases because of the static hardening contribution of the dynamically formed hcp phase. Thus, the decrease of the strain hardening rate starts to slow down at a stress that depends on initial grain size and deformation mode (Figure 9). The slower decrease of the strain hardening rate at larger strains is important because it helps to maintain stable, uniform flow until the applied stress reaches the magnitude of the fracture strength.

Effect of Carbon Content on Strain Hardening - The effects of flow stress and cooling rate on the strain hardening rate of alloys C1 and C3 are shown in (Figure 11). The highest strain hardening rates are observed from the proportional limit (300 MPa for the present alloys) to a stress that increases with C content: 350 and 490 MPa for the alloys with 0.05 and 0.31 wt% C, respectively. At higher stresses, the strain hardening rate exhibits a stronger dependence on flow stress and C content. In the case of the 0.31 wt% C alloy, the yield strength and the rate of decrease of θ with flow stress ($\delta\theta/\delta\sigma$) are not affected significantly by the cooling rate. It is noteworthy that transition region from high to low $\delta\theta/\delta\sigma$ takes place smoothly between 700 and 850 MPa. A similar behavior is observed in the furnace cooled, 0.05 wt% C alloy, although the transition region takes place at lower stresses (Figure 11a). In contrast, the water quenched, 0.05 wt%C alloy exhibits a significantly faster decrease of θ with increasing flow stress (Figure 11a) and the transition region takes place rapidly at about 500 MPa. In this case, however, the region of rapid change of θ with stress leads to a region where the θ remains constant until the material fractures. This behavior is similar to that observed for the annealed Co-27Cr-5Mo-0.05C wrought alloys (Figure 9).

The main microstructural difference between high C alloys, the furnace cooled low C alloy and the water quenched, low C alloy (Figures 3 and 7) is that the first three materials contain a significantly higher volume fraction of second phase carbide particles. During the initial stages of plastic deformation the flow stress depends strongly on the rate of strain hardening. Thus, the data in (Figure 11) suggest that the presence of second phase particles in the microstructure causes a rapid increase of the flow stress for a given increment of plastic strain which leads to a slower decrease of the strain hardening rate with increasing strain or stress. The faster decrease in θ at low stresses and the region of constant θ at high stresses observed in the water quenched, 0.05 wt% C alloy (Figure 11a), suggest that strain-induced fcc(metastable)=>hcp phase transformation contributes significantly to facilitate stable plastic deformation. The X-ray diffraction patterns presented in (Figure 8) support this argument. It can therefore be concluded that the strain hardening behavior in high C, solution treated Co-27Cr-5Mo-C alloys is controlled by the interaction between strain-induced crystal defects and second phase carbide particles. In contrast, the strain-induced fcc-hcp phase transformation plays a major role in controlling the plastic deformation and hardening behaviors of water quenched, low C alloys.

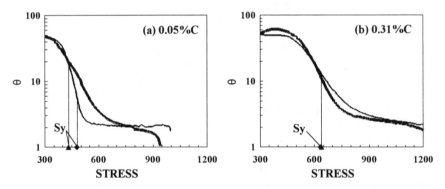

Figure 11-*Effects of cooling rate (thin line: water quenching, thick line: furnace cooling) on the stress denpendence of the strain hardening rate of cobalt-base alloys solution treated at 1250 °C.*

Conclusions

Strain-induced fcc(metastable)=>hcp phase transformation plays an important role during the large strain plastic deformation of Co-Cr-Mo-C alloys. The microstructures and the strain hardening behavior observed indicate that the dynamic formation of the hcp phase acts as a soft strain producing mechanism at low stresses and causes a rapid decrease in the strain hardening rate. As the flow stress increases, the rate of decrease of the strain hardening rate decreases gradually leading to regime of constant hardening rate that starts at a stress that depends on initial grain size and deformation mode. This behavior implies that the larger strain hardening rate required to maintain uniform flow at high stresses can be attributed to the static hardening contribution of the dynamically formed hcp phase. Under these conditions, uniform plastic flow will continue until the applied stress reaches the magnitude of the fracture strength of the material.

In solution treated cast alloys, the flow and hardening behaviors depend strongly on carbon content. It was shown that high carbon contents inhibit the occurrence of strain-induced phase transformation and the flow and hardening behavior of these materials depend on the interactions between strain-induced crystal defects and second phase carbide particles present in the microstructure.

Acknowledgments

The author is indebted to the National Science and Technology Council of Mexico for the financial support of this work through Projects 3076A and 26320-A.

References

[1] Vander Sande, J. B., Coke, J. R., and Wulf J., "Transmission Electron Microscopy Study of The Mechanisms of Strengthening in Heat Treated Co-Cr-Mo-C Alloys," *Metallurgical Transactions A*, Vol. 7A, 1976, pp. 389-397.

[2] Rajan, K., and Vander Sande, J. B., "Room Temperature Strengthening Mechanisms a Co-Cr-Mo-C Alloy," *Journal of Materials Science*, Vol. 17, 1982, pp. 769-778.

[3] Rajan, K., "Phase Transformations in a Wrought Co-Cr-Mo-C Alloy," *Metallurgical Transactions A*, Vol. 13A, 1982, pp. 1161-1166.

[4] Kilner, T., Laanemäe, W. M., Pilliar R., Weatherly, G.C., and MacEwen, S. R., "Static Mechanical Properties of Cast and Sinter-Annealed Cobalt-Chromium Surgical Implants," *Journal of Materials Science*, Vol. 21, 1986, pp. 1349-1356.

[5] Saldivar, G. A., "*In-Situ X-Ray Diffraction Study of the Isothermal Martensitic Transformation in Co-27Cr-5Mo-0.05C Alloy*," Ph. D. Thesis, CINVESTAV, Saltillo, Mexico, 1998.

[6] Graham, A., and Youngblood, J. L., "Work Strengthening by a Deformation-Induced Phase Transformation in MP Alloys," *Metallurgical Transactions*, Vol. 1, 1970, pp. 423-430.

[7] Olson, G. B., and Cohen, M., "Kinetics of Strain-Induced Martensitic Nucleation," *Metallurgical Transactions A*, Vol. 6A, 1975, pp. 791-795.

[8] Salinas-Rodríguez, A., and Rodriguez-Galicia, J. L., "Deformation Behavior of Low-Carbon C0-Cr-Mo Alloys for Low-Friction Implant Applications," *Journal of Biomedical Materials Research*, Vol. 31, 1996, pp. 409-419.

[9] Kilner T., Weatherly G. C., and Pilliar R. M., "Morphological Studies of Incipient Melting Phenomena in a Cobalt-Based Alloy," *Scripta Metallurgica*, Vol. 16, 1982, pp. 741-744.

[10] Sage, M., and Gillaud, Ch., "Méthode d'analyse quantitative des variétés allotropiques du cobalt par les rayons X," *Revue de Metallurgie*, Vol. 47, 1950, pp. 139-144.

Wear Characterization

Adi Wang,[1] J. Dennis Bobyn,[2] Steve Yue,[1] John B. Medley,[3] and Frank W. Chan[2]

Residual Abrasive Material from Surface Grinding of Metal-Metal Hip Implants: A Source of Third-Body Wear?

Reference: Wang, A., Bobyn, J. D., Yue, S., Medley, J. B., and Chan, F. W., "**Residual Abrasive Material from Surface Grinding of Metal-Metal Hip Implants: A Source of Third-Body Wear?**", *Cobalt-Base Alloys for Biomedical Applications, ASTM STP 1365*, J. A. Disegi, R. L. Kennedy, and R. Pilliar, Eds., American Society for Testing and Materials, West Conshohocken, PA, 1999.

Abstract: The surfaces of fifteen heads of all metal hip implants were investigated using a scanning electron microscope. The heads were manufactured from Co-Cr-Mo alloy, either cast, low carbon wrought or high carbon wrought. Six of the surfaces were investigated prior to hip simulator testing, three after 3 million cycles of simulator testing and six after 6 million cycles of testing. Third-body wear type scratches of varying amounts were observed on all fifteen heads. Residual grinding stone material from the manufacturing process was also found at varying amounts on the surfaces of all fifteen heads. There was a correlation between the extent of this residual material and the extent to which the surfaces were scratched.

Keywords: hip implant, metal-on-metal, articulating surface.

Introduction

Metal-metal bearing surfaces made from cobalt-chromium-molybdenum (Co-Cr-Mo) alloy are being increasingly used in artificial hip replacement because of their excellent wear resistance. After the original metal-metal experience in the 1960's with devices such as the McKee-Farrar, a second generation of implants with improved alloy properties and processing techniques has evolved over the past ten years [1-6]. The surface finish of the implants, a factor shown to affect wear performance [1,2], has been improved by more sophisticated grinding and polishing technologies. However, modern implants retrieved from patients and those tested in hip simulators are commonly characterized by numerous surface scratches of greater width and depth than those created during manufacturing [1,2,4-8]. It has been proposed that the scratches are caused by 3rd-body abrasion of running-in material [9] produced during the initial period

---

[1]B.Eng. (M.Eng. candidate) & Associate Professor, respectively, Department of Metallurgical Engineering, McGill University, Montreal, Quebec, Canada.

[2]Associate Professor & M.Eng. (Ph.D. candidate), respectively, Jo Miller Orthopaedic Research Laboratory, Montreal General Hospital, Montreal, Quebec, Canada.

[3]Associate Professor, Mechanical Engineering Department, University of Waterloo, Waterloo, Ontario, Canada.

of implant use or by conventional abrasion from protruding hard carbides [5,6,8] that exist within the cast and high carbon wrought alloys. The purpose of the present study was to analyze the surfaces of Co-Cr-Mo alloy heads before and after simulator testing in an attempt to ascertain the mechanism(s) responsible for the abrasive scratching. The present study concentrated exclusively on this issue whereas a more general characterization of the surfaces was performed in an ancillary investigation [10].

Materials and Methods

The head components were manufactured from the following Co-Cr-Mo alloys: cast ASTM F75-92, low carbon (LC) wrought ASTM F1537-94 or high carbon (HC) wrought ASTM F1537-94. They were produced by a medical device manufacturer using state-of-the-art fabrication techniques. After numerical control machining, the heads were ground to size (28 mm diameter) with roughing stones and finishing stones employed in a rotary fashion, and finally polished with diamond paste. This procedure was a standard superfinishing process for the manufacture of metal-metal hip implants. Final cleaning of the heads was accomplished by ultrasonic immersion in sodium borate and sodium hydroxide solution baths, followed by rinsing in water. The components were subsequently passivated according to standard ASTM procedures.

The grinding stones used during manufacturing are commonly utilized for metal-metal implant fabrication and were made of SiC abrasives that were bonded with a silica-alumina glass. They were manufactured by mixing SiC abrasive particles (of a known average size) with a small amount of bond pellets. The desired stone shape was prepared by taking this mixture and pressing it into a mold. The pressed shape was sintered (in a high temperature furnace) to allow the bond pellets to soften and flow around the SiC abrasives in order to consolidate the abrasives. The roughing stones contained SiC abrasives that were several microns in diameter, and were about one order of magnitude larger than the abrasives of the finishing stones. The bond pellets that were mixed with the SiC abrasives were of a size between the roughing and finishing abrasives.

For this study, the articulating surfaces of fifteen implant heads were examined. Six of the heads (two of each alloy) were examined as manufactured (i.e., without any hip simulator testing). Three heads (one of each alloy) had been tested in a model EW08 MMED hip simulator[4] for 3 million cycles, and six heads (3 of the cast and 3 of the HC wrought) had been tested in the hip simulator for 6 million cycles. The hip simulator subjected the articulating surfaces to a biaxial rocking motion with a peak load of 2100N applied in a gait-type loading cycle at a rate of 1.13Hz. Testing methods and conditions are described in detail by Chan et al. [1,2]. Information on the specimens is summarized in Table 1, along with the gravimetrically obtained volumetric wear resulting from · simulator testing performed by Chan [11].

The articulating surfaces of the heads were examined using a Jeol 840 scanning electron microscope (SEM) at accelerating voltages of 5 or 10keV. The SEM was used for topographical and compositional imaging as well as for qualitative elemental analysis using a NORAN I2 energy dispersive spectrometer (EDS) that was equipped with an ultra thin window for detection of low atomic number elements. The surfaces were prepared for examination by cleaning with acetone and/or replica tape.

[4] MATCO, La Canada, CA.

Sections of roughing and finishing grinding stones from the manufacturing process were available for examination. They were analyzed with the SEM topographically, and the EDS was also used. Random fragments from the grinding stones were mounted on studs with carbon tape and were sputter coated with gold prior to SEM examination.

Table 1 – *The Volumetric Wear of Each Implant*

Alloy	Cycles of testing	Volumetric wear, mm^3
Cast	0	0
Cast	0	0
LC wrought	0	0
LC wrought	0	0
HC wrought	0	0
HC wrought	0	0
Cast	3 million	0.159
LC wrought	3 million	0.960
HC wrought	3 million	0.150
Cast	6 million	0.700
Cast	6 million	0.610
Cast	6 million	0.780
HC wrought	6 million	0.610
HC wrought	6 million	1.030
HC wrought	6 million	0.700

Results

3rd-Body Scratches

The surfaces of the untested heads were generally characterized by randomly-oriented shallow surface scratches resulting from the grinding and polishing phases of manufacturing (Figure 1). All 15 heads, including the six untested heads, had regions in the wear zone of deeper, more aggressive scratches that were suggestive of 3rd-body abrasive wear (Figure 2). Thus, for the purposes of discussion, these scratches were referred to as "3rd-body scratches". From a qualitative standpoint, the extent of these 3rd-body scratches was greater with the cast components and lesser with the HC wrought components. The extent of 3rd-body scratches on the cast components was greatest at 3 million cycles of testing, but decreased between 3 and 6 million test cycles. The extent of 3rd-body scratches on the LC wrought components was also greatest at 3 million cycles of testing when compared to before testing (LC wrought heads were not available for examination after 6 million cycles). By contrast, very few 3rd-body scratches were observed on any of the HC wrought components regardless of the number of test cycles.

Figure 1 - *SEM image of shallow scratches on implant surface resulting from grinding and polishing during manufacturing. Implant has not yet been tested in the simulator.*

Figure 2 - *SEM image of 3rd-body abrasive scratches on cast component after 3 million test cycles. These scratches were not typical of the entire wear zone, but rather were found in isolated areas.*

Two main types of 3rd-body scratches were observed on the implants: some were straight while others were tortuous or irregular in shape. These two types were seen both individually in different regions as well as together in the same region (Figure 2). Two of the heads exhibited straight, near parallel scratches that met at an origin (Figure 3).

Figure 3 - *SEM image of 3rd-body abrasive wear on HC wrought component after 6 million test cycles. Image in window shows the origin where the scratches meet.*

Residual Stones on Articulating Surfaces

All fifteen heads possessed small embedded fragments of grinding stone material on their surfaces, henceforth referred to as "stones", that appeared to have been left behind from the manufacturing process. The stones were mostly spherical (Figure 4), but some were irregular in shape. They ranged from about 1 to 5μm in diameter or greater, and varied in the depth to which they were embedded.

The stones were characterized by EDS (Figure 5) and were shown to be composed of SiC, SiO_2-Al_2O_3 (glass) and MgO. The relative amount of these constituents varied from stone to stone. Most often, only one or two of the constituents were present. The irregularly shaped stones tended to have more SiC, while the spherical stones had mainly SiO_2 and Al_2O_3.

The HC wrought components possessed very few stones prior to testing and at 3 and 6 million cycles. The LC wrought components contained more stones prior to testing and showed a decrease in their prevalence at 3 million cycles. Of the three alloys, the cast components had the greatest amount of residual stones prior to testing and showed a progressive decrease at 3 and 6 million cycles. The stones remaining on the surfaces after 6 million cycles tended to be of the larger size.

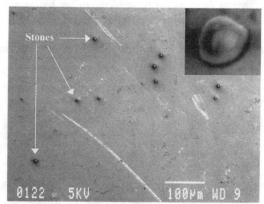

Figure 4   -   *SEM image of spherical stones embedded in the surface of a cast implant which has not yet been tested.   Image in window is a close-up of a single stone.*

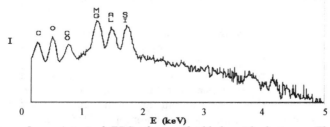

Figure 5   -   *A typical EDS of an embedded residual stone.   The scan showed the stone to be likely comprised of SiC, SiO_2, Al_2O_3 and MgO.*

Analysis of Grinding Stones Used During Manufacturing

The discovery of residual grinding stone material on the implant surfaces prompted an investigation of the grinding stones used for manufacturing. In general, the roughing and finishing grinding stones exhibited similar morphologies (Figure 6). The main difference was in the size of the abrasives; the finishing having about an order of magnitude finer abrasives than the roughing, which were several micrometers in diameter. The shape of the abrasives was similar to that of the irregularly shaped residual stones found on the implants.

Occasionally, pockets of spherical particles were found in the roughing grinding stone (Figure 7). The spherical particles possessed the same morphology and general size as the spherical stones found on the implants (Figure 4). EDS revealed that the spherical particles in the roughing grinding stones were mainly composed of SiO_2 with Al_2O_3; the same result was obtained when the spherical stones on the implants were analyzed. Because of their composition, these particles were identified as bond pellets. Their morphology and isolated nature suggested that they did not fully consolidate with the abrasive during sintering. These unconsolidated bond pockets were not found in the finishing grinding stones. To examine this issue more closely, two additional cast alloy implants were analyzed. One of these was only subjected to machining and grinding with the roughing grinding stone, while the other was further ground with the finishing grinding stone (neither one was polished). SEM examination of these two implants revealed that the residual stones were primarily embedded in the implant surfaces during the rough grinding step.

(a)   (b)

Figure 6 - *SEM images of fragments of grinding stones used during implant manufacturing. a) roughing stone, b) finishing stone.*

A scratch test using consolidated bond pellets (without SiC) was conducted in order to determine if the bond material was hard enough to be able to abrade the implant alloys. The test was done by taking a fragment from the consolidated bond and scratching the surfaces of the implants, by hand, with the bond fragment. All three alloys were easily scratched with light manual pressure (Figure 8). Surrounding the scratched area were many fragments of the bond material (Figure 8b).

Figure 7 - *SEM image of a pocket of spherical bond pellets in the roughing stone used during manufacturing. Arrow points to a pellet with the same appearance as that shown in the insert in Figure 4.*

(a)   (b)

Figure 8 - *SEM images of a surface which was manually scratched with bond material. a) before scratch test, b) after scratching. The window in (b) is a close-up of the scratched area which shows bond fragments embedded in the surface.*

Discussion

Several authors [1,2,4-9] had previously reported scratches that seemed similar to the ones observed in the present study. Schmidt et al. [5], Streicher et al. [6] and Park et al. [8] have proposed that the scratches were caused by carbide fragments that were released from the implant surface during articulation. Walker and Gold [9] were less specific bud did suggest that the scratches were caused in part by wear particles formed during the running-in period of wearing.

While these scratches might have occurred as previously proposed, the finding of residual grinding stone material on the implant surfaces strongly indicated an additional mechanism. This was supported by the finding that the extent of scratching correlated directly with the amount of residual grinding material on the component surfaces for all alloys. Furthermore, the widths of the scratches were compatible with the stone sizes.

The extent of this 3rd-body scratching was greatest at 3 million cycles and decreased between 3 and 6 million cycles of testing. An ancillary study [10] which examined the same components revealed a small amount of carbide release from the surface during the first 3 million cycles of testing but relatively more carbide release between 3 and 6 million cycles. If carbides were the main cause of the 3rd-body type scratching, it would be expected that an increase in carbide release would result in an increase in scratches, but this was not the case in the present study. However, Park et al. [8] found that the surface scratches seemed to be caused by carbides that were released during the initial running-in period and after running-in, the carbide release apparently ceased, and correspondingly, so did the scratching. In the present study, it was important to note that scratches were readily seen in the LC wrought head that did not have large enough carbides [10] to cause such wide scratches. Also, the LC wrought displayed more scratches than the HC wrought even though the HC wrought had a far greater amount of carbides. These findings together implied that the carbides were not the main cause of the 3rd-body type scratching.

There were two main types of stones found on the surfaces: irregularly shaped stones made mostly of SiC and spherical stones made mostly of SiO_2 and Al_2O_3. It was clear that the SiC stones were capable of scratching the implant surfaces, since SiC is the abrasive of choice for grinding during manufacturing. However, the SiC stones were found in far fewer number than the SiO_2-Al_2O_3 bond stones. Therefore, it was important to determine if the bond was capable of scratching the implant alloys. The scratch tests performed with the bond material (Figure 8) clearly showed that the bond was hard enough in itself to abrade the alloys. Furthermore, this test also showed the ease with which the bond can imbed itself in the implant material (Figure 8b).

The 3rd-body type scratches that were observed on the component surfaces were either in straight lines or tortuous. The straight-line scratches might reasonably be expected to be caused by embedded stones on an opposing surface. The tortuous scratches might result from stones that were released from one of the articulating surfaces giving them freedom to abrade in random directions. It was clear that the stones were capable of being dislodged from the surfaces because a decrease in the number of stones was observed with an increase in test cycles.

The cast implants consistently displayed more residual grinding stones than the wrought heads. This might reflect differences in surface hardness between the alloys. It may also be related to the coarser grain structure and carbide size of the cast material, factors known to render machining and finishing more difficult than with finer structures. The high carbon wrought alloy appeared more resistant to the embedding of residual grinding stones; this might be an important manufacturing consideration.

The present study only examined the abrasive scratching of the simulator tested surfaces. An ancillary study by Wang et al. [10] reported a more general surface characterization of the same implant heads. Also, the present study did not correlate the extent of 3rd-body scratches and residual grinding stone material with wear performance of the different alloys. Studies by Chan et al. [1,2] have shown no significant differences between the overall wear of the three alloys after 3 million cycles of testing. This suggested that the presence of residual grinding stones did not have a large influence on global wear performance. However, more systematic microscopic and tribological studies on a larger number of implants would be required to definitely ascertain the magnitude of this influence. Apart from any issues related to the cellular response to foreign material, the wear performance might possibly be improved in the absence of residual grinding stones. This microscopic analysis revealed a hitherto unreported source of 3rd-body scratching on Co-Cr-Mo alloy self-bearing hip implants. This raises awareness of the need for more careful cleaning after manufacturing or modification of the grinding and polishing phases during manufacturing. It is clear, however, that once residual grinding stone particles have been embedded in the metal implant surfaces, they are very difficult to remove even when diligent cleaning is employed. Studies such as this should be extended to components from different manufacturers to characterize the cleanliness of other implants destined for clinical use.

Acknowledgements   Medical Research Council of Canada, Natural Science and Engineering Research Council of Canada.

References

[1]   Chan, F. W., Bobyn, J. D., Medley, J. B., Krygier, J. J., Yue, S. and Tanzer, M., "Engineering Issues and Wear Performance of Metal on Metal Hip Implants", *Clinical Orthopaedics and Related Research*, Vol. 333, Dec. 1996, pp. 96-107.

[2]   Chan, F. W., Bobyn, J. D., Medley, J. B., Krygier, J. J., Podgorsak, G. F. and Tanzer, M., "Metal-Metal Hip Implants: Investigation of Design Parameters that Control Wear", *Clinical Orthopaedics and Related Research*, 1999, (in press).

[3]   Medley, J. B., Chan, F. W., Krygier, J. J. and Bobyn, J. D., "Comparison of Alloys and Designs in a Hip Simulator Study of Metal on Metal Implants", *Clinical Orthopaedics and Related Research*, Vol. 329S, Aug. 1996, pp. S148-S159.

[4]   Rieker, C. B., Kottig, P., Schon, R., Windler, M. and Wyss, U. P., "Clinical Wear Performance of Metal-on-Metal Hip Arthroplasties", *Alternative Bearing Surfaces in Total Joint Replacement, ASTM STP 1346*, J. J. Jacobs and T. L. Craig, Eds., American Society for Testing and Materials, 1998.

[5]   Schmidt, M., Weber, H. and Schon, R., "Cobalt Chromium Molybdenum Metal Combination for Modular Hip Prostheses", *Clinical Orthopaedics and Related Research*, Vol. 329S, Aug. 1996, pp. S35-S47.

[6]   Streicher, R. M., Semlitsch, M., Schon, R., Weber, H. and Rieker, C., "Metal-on-Metal Articulation for Artificial Hip Joints: Laboratory Study and Clinical Results", *Journal of Engineering in Medicine*, vol. 210, No. 3, 1996, pp. 223-232.

[7]   McKellop, H., Park, S.-H., Chiesa, R., Doorn, P., Lu, B., Normand, P., Grigoris, P. and Amstutz, H., "In Vivo Wear of 3 Types of Metal on Metal Hip Prostheses During 2 Decades of Use", *Clinical Orthpaedics and Related Research*, Vol. 329S, Aug. 1996, pp. S128-S140.

[8]   Park, S.-H., McKellop, H., Lu, B., Chan, F. and Chiesa, R., "Wear Morphology of Metal-Metal Implants: Hip Simulator Tests Compared with Clinical Retrievals", *Alternative Bearing Surfaces in Total Joint Replacement, ASTM STP 1346*, J. J. Jacobs and T. L. Craig, Eds., American Society for Testing and Materials, 1998.

[9]   Walker, P. S. and Gold, B. L., "The Tribology (Friction, Lubrication and Wear) of All-Metal Artificial Hip Joints", *Wear*, Vol. 17, 1971, pp. 285-299.

[10]   Wang, A., Yue, S., Bobyn, J. D., Chan, F. W. and Medley, J. B., "Surface Characterization of Metal-on-Metal Hip Implants Tested in a Hip Simulator", *Wear*, 1999, (in press).

[11]   Chan, F. W., "Wear and Lubrication of Metal-Metal Bearings for Total Hip Arthroplasty", *Ph.D. Thesis*, Department of Biomedical Engineering, McGill University, 1999.

Kathy K. Wang,[1] Aiguo Wang,[1] and Larry J. Gustavson[1]

Metal-on-Metal Wear Testing of Co-Cr Alloys

Reference: Wang, K. K., Wang, A., and Gustavson, L. J., **"Metal-on-Metal Wear Testing of Co-Cr Alloys,"** *Cobalt-Base Alloys for Biomedical Applications, ASTM STP 1365,* J. A. Disegi, R. L. Kennedy, and R. Pilliar, Eds., American Society for Testing and Materials, West Conshohocken, PA, 1999.

Abstract: The wear characteristics of various Co-Cr alloy combinations were studied using a reciprocating wear machine. Wear test specimens were made from three Co-Cr alloys, cast Co-Cr, low carbon wrought Co-Cr and high carbon wrought Co-Cr alloys. The cast Co-Cr alloy was evaluated in both the as-cast and the solution-treated conditions. All specimens were polished with a surface roughness in the range of 0.01-0.02 μm. The clearance in diameter between the convex and the concave specimens was 100 or 300 μm. The same test conditions were applied to all specimens. Results showed that the best alloy couple was as-cast Co-Cr on as-cast Co-Cr alloy and this couple was found to be superior to the high carbon wrought on the high carbon wrought couple. This finding is also supported by using a hip simulator wear test machine.

Keywords: Co-Cr alloys, metal-on-metal wear, wear resistance, reciprocating wear testing, surface roughness

Metal-on-metal total hip prostheses made of the cast Co-Cr alloy were used for hip replacement in the 60s and 70s, but by the mid-70s they were replaced by polyethylene-on-metal bearings. Main factors that led to the abandonment of the metal-on-metal bearings were the early success of the Charnley prosthesis [1,2], and the seizing and cup loosening problems experienced with metal-on-metal prosthesis [3-8]. However, a number of metal-on-metal hip implants have survived for more than 25 years with low wear rates and minimal osteolysis [9]. This fact suggests that with correct design metal-on-metal systems can work.

---

[1] Assistant Director, Assistant Director, and Director, respectively, Research and Development, Howmedica Inc., Pfizer Medical Technology Group, 309 Veterans Blvd., Rutherford, NJ 07070.

The encouraging long-term results have aroused interest since the mid-80s in developing a new generation of metal-on-metal articulations [10,11]. The purpose of this work was to determine the wear characteristics of various Co-Cr alloy combinations using a reciprocating wear machine. The best two alloy combinations were then tested using a hip simulator wear test machine.

Materials

Three commercially available Co-Cr alloys, the cast Co-Cr, low carbon wrought Co-Cr and high carbon wrought Co-Cr alloy, were selected to be studied. The characteristics of each alloy are briefly described as below.

Both the as-cast and the solution-treated (S.T.) cast Co-Cr alloy which meet the requirements of ASTM F-75, Standard Specification for Cast Cobalt-Chromium-Molybdenum Alloy for Surgical Implant Application, were studied. The cast Co-Cr alloy was used in the old McKee-Farrar system that could be made from either as-cast or solution treated condition. The typical carbon content for cast Co-Cr alloy is 0.25 wt.%. The cast Co-Cr alloy has a coarse dendritic structure characteristic of investment casting and is strengthened by the presence of carbides. The solution treatment used for cast Co-Cr alloy serves to homogenize the cast structure and was conducted at 1218°C for one hour. The hardness of as-cast Co-Cr alloy is 31-33 Rc. That of solution treated cast alloy is 28-29 Rc. Typical grain structures for the as-cast and the solution treated cast Co-Cr alloys are shown in Figures 1 and 2 respectively.

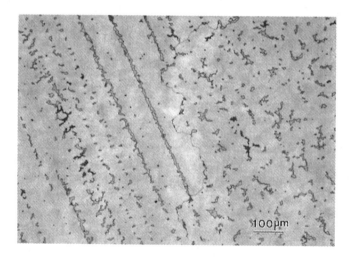

Fig. 1—*A typical grain structure of as-cast Co-Cr alloy.*

The low carbon wrought Co-Cr alloy meets the requirements of ASTM F-1537, Standard Specification for wrought Cobalt-28 Chromium-6 Molybdenum Alloy for Surgical Implants and has a typical carbon content of 0.04-0.05 wt.%. The material used in this study is in the as-rolled condition and has a hardness of 40 Rc. This material is strengthened by fine grain structure through thermal mechanical processing.   A typical grain structure of the low carbon wrought Co-Cr is shown in Figure 3.

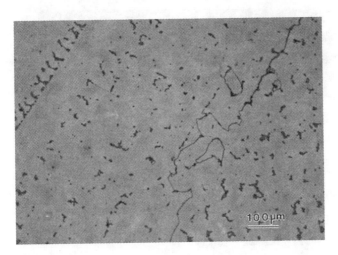

Fig. 2 — *A typical grain structure of solution treated cast Co-Cr alloy.*

Fig.3 — *A typical grain structure of low carbon wrought Co-Cr alloy.*

The high carbon wrought Co-Cr alloy meets the requirements of ASTM F-1537 and has a typical carbon content of 0.20-0.25 wt.%. The bar material used in this study was in the as-rolled condition and made by ingot metallurgy. It exhibits fine chromium carbides, with a hardness of 43 Rc. The carbide particles present in the alloy are less than 5 µm and much smaller than those in the cast Co-Cr alloy. A typical grain structure is shown in Figure 4.

Fig. 4 — *A typical grain structure of high carbon wrought Co-Cr alloy.*

Test Specimens

Eight different alloy combinations (Table 1) were studied to determine the wear characteristics of three different Co-Cr alloys. The cast specimens were investment cast at Howmedica. Specimens other than the cast ones were machined from bar stocks of the above mentioned alloys. The convex specimen has a nominal radius of curvature of 36.02 or 35.98 mm and a width of 17.8 mm. The concave specimen has a nominal radius of curvature of 36.07 or 36.13 mm and a width of 12.7 mm. The diametral clearance between the convex and the concave specimens was 100±50 µm or 300±50 µm. The average surface roughness (R_a) was 0.01- 0.02 µm. All test specimens were cleaned and passivated prior to testing. The number of pairs tested for each alloy combination is shown in Table 1.

Test Method and Procedures

The test apparatus used in this study was a 12-station reciprocating wear machine. A schematic test setup for reciprocating testing is shown in Figure 5. A constant load of 500 N was applied to each station during testing. Filtered bovine serum (HyClone, Inc.) with 0.2 wt.% sodium azide was used as lubricant. All tests were conducted at room temperature. To prevent excessive frictional heat build-up in the specimen chamber, the serum was constantly circulated between each chamber and a reservoir connected to it by a pump. During the test, the convex surface oscillated on the concave surface at a frequency of 1 Hz. The amplitude of oscillation was +/- 30° that gave rise to a sliding distance of 37.62 mm per cycle. All tests were run for 250,000 cycles. The wear of the specimens was quantified by weight loss measurement using an electronic analytical balance with a measurement resolution of ±0.01 mg (Denver Instrument Company, Model AB-250D).

Table 1 – *Test Co-Cr Specimens*

Alloy Combnation	No. of Alloy pairs	Convex Specimen	Concave Specimen
1^1	6	As-cast	As-cast
2^1	6	S.T. cast	S.T. cast
3^1	6	S.T. cast	As-cast
4^1	4	Low carbon wrought	Low carbon wrought
5^1	4	Low carbon wrought	S.T. cast
6^1	4	High carbon wrought	High carbon wrought
7^1	3	High carbon wrought	S.T. cast
8^1	1	High carbon wrought	As-cast
1^2	3	As-cast	As-cast
6^2	3	High carbon wrought	High carbon wrought
8^2	3	High carbon wrought	As-cast

^{1} Specimens having a clearance of 100±50 μm.
^{2} Specimens having a clearance of 300±50 μm.

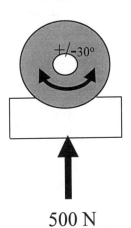

500 N

Fig. 5—*Schematic setup for the reciprocating wear test.*

Results and Discussion:

The average weight loss for each specimen with 100±50 μm diametral clearance and for each combination is shown in Table 2. Results show that the low carbon wrought alloy and the solution treated cast Co-Cr alloy had very high wear rates and are not suitable for metal-on-metal applications. The high carbon wrought Co-Cr alloy (convex) coupled with as-cast Co-Cr alloy (concave) seems to be a better combination than the as-cast on as-cast couple. Unfortunately only one pair of wrought/cast specimens was tested.

Since there was a wide variation (large standard deviations) in the weight loss for most alloy combinations, it was suspected that the diametral clearance (100±50 μm) may be too small for the 72 mm diameter specimens. For this reason, tests were conducted on the three promising alloy combinations, as-cast vs. as-cast, high carbon wrought vs. as-cast, and high carbon wrought vs. high carbon wrought. The clearance of these specimens was increased to 300±50 μm. Three pairs for each alloy combination were tested.

The average weight loss after 250,000 cycles for each combination is shown in Table 3. Results showed that the high carbon wrought coupled with high carbon wrought alloy had much higher wear than the other two alloy combinations (as-cast vs. as-cast and high carbon wrought vs. as-cast). Since the as-cast/as-cast and the high carbon wrought/as-cast alloy combination showed essentially no wear after 250,000 cycles, these two alloy combinations were cleaned and tested to 500,000 cycles. The average weight loss after 500,000 cycles is shown in Table 4. As can be seen in Table 4, the average weight loss of as-cast/as-cast alloy combination is lower than that of high carbon wrought/as-cast alloy combination after 500,000 cycles. The high standard deviation was resulted from one test pair from each alloy combination had a high wear after 500,000 cycles possibly due to misalignment during testing.

Table 2 — *Average Weight Loss (mg) and Standard Deviation Data after 250,000 Cycles for the Specimens with 100±50 μm Diametral Clearance[1]*

Convex/Concave	Convex Wear	Concave Wear	Total Wear
As-cast/As-cast	4.1 ± 4.0	65 ± 45	69.1 ± 49.0
S.T. cast/S.T. cast	37 ± 31	114 ± 34	150 ± 51
S.T. cast/As-cast	3.1 ± 0.7	102 ± 12	105.1 ± 12.7
Low carbon/Low carbon	471 ± 51	187± 28	658 ± 77
Low carbon/S.T. cast	470 ± 54	29 ± 5	499 ± 57
High carbon/High carbon	69 ± 60	85 ± 74	154 ±134
High carbon/S.T. cast	86 ± 81	68 ± 40	154 ± 121
High carbon/As-cast	0	0.02	0.02

[1] The number of pairs for each alloy combination can be found in Table 1.

Table 3 — *Average Weight Loss (mg) and Standard Deviation Data after 250,000 Cycles for the Specimens with 300±50 µm Diametral Clearance, N=3*

Convex/Concave	Convex Wear	Concave Wear	Total Wear
As-cast/As-cast	-0.2 ± 0.3	0.4 ± 0.2	0.2 ± 0.1
High carbon/High carbon	152.9 ± 27.6	146.6 ± 7.9	299.6 ± 35.0
High carbon/As-cast	10.1 ± 13..3	53.5 ± 61.7	63.5 ± 75.0

As compared to the wear data shown in Table 2, it was found that the weight loss of larger diametral clearance (300±50 µm) between the test specimens was much lower than that with smaller clearance for the as-cast/as-cast alloy combination. Opposite findings were found for the wrought alloy combination.

Table 4 — *Average Weight Loss (mg) and Standard Deviation Data after 500,000 Cycles for the Specimens with 300±50 µm Diametral Clearance, N= 3*

Convex/Concave	Convex Wear	Concave Wear	Total Wear
As-cast/As-cast	2.0 ± 4.1	23.1 ± 38.4	25.1 ± 42.4
High carbon/As-cast	6.8 ± 10.9	38.1 ± 65.4	44.8 ± 76.1

Since there is still a wide variation in the weight loss for the as-cast/as-cast and high carbon wrought/as-cast alloy combination with large clearance, it was decided to conduct wear testing on the as-cast/as-cast combination using the MTS hip joint simulator to verify the reciprocating test results. The 28 mm cast cups and heads were investment cast at Howmedica. The diametral clearance between the cup and head was 100 µm. The average surface roughness (R_a) of the cups was 0.035 µm. That of the heads was 0.025 µm. The sphericity was controlled to be less than 7 µm.

Configuration of a hip simulator station is shown in Figure 6. Each station consists of fixtures for accommodating the head and the insert which are aligned along the same vertical axis. An acrylic specimen chamber with a stainless steel baseplate is used. The spigot has an appropriate taper that accommodated the head. The specimen chamber fits onto a block inclined at 23°. The head and the insert were immersed in bovine calf serum during testing, which was contained in an acrylic chamber. Testing was conducted in the anatomical orientation, in which the insert is in the superior position and is stationary, the head articulates inside the insert. The inclined block rotates at a speed of 1 Hz and compressive loading was applied axially with a maximum load of 2450 N. The loading is cyclic in nature and follows the physiological profile determined by Paul [12]. The components were tested in filtered bovine serum with the addition of ethylenediamine-tetraacetic acid (EDTA). The serum was replaced every 0.25 million cycles. Head and cup components were weighed at the same intervals. Tests were run up to one million cycles.

Sulzer 28 mm Metasul® components were also tested as controls. Sulzer Metasul® components were chosen because they represent the new generation of metal-on-metal products and made of high carbon wrought Co-Cr alloy. The dimension (28 mm), surface roughness (0.01-0.04 μm), sphericity (<7 μm), diametral clearance (100 μm) and design of the as-cast cup and head components were comparable to those of Sulzer components.

Table 5 shows the run-in wear volume losses of the cups, the heads, and the total wear for the Sulzer Metasul® and experimental cast components. Among the four Sulzer cup/head paired components, only one pair showed a high total wear rate (>13 mm^3/million cycles) while the rest of the components exhibits very low wear rates (<6 mm^3/million cycles). The two experimental cup/head samples showed very low wear rate, comparable to that of the three low wear Sulzer components. The pair that showed high wear rate could be due to the frequent mounting and dismounting of the head/cup component at every 0.25 million cycles which may caused misalignment.

Excluding the Sulzer cup/head component #3, the other five components have a run-in wear volume loss of 1.713 to 5.740 mm^3/million cycles that is comparable to the published run-in wear volume loss data [13].

Fig. 6 — *Configuration for each station of the hip simulator.*

Considering the clinical wear behavior of the McKee-Farrar prostheses which either failed early on during the first few years of implantation or survived for up to 30 years [9], it may be more appropriate to use the running-in wear data (one-million cycles or less) as a more reliable measure of the true wear behavior of the metal-on-metal couples from the hip simulator test. The hip simulator wear tests in this study were stopped at 1 million cycles to determine the initial run-in wear rates.

Table 5 — *Initial Period of Accelerated Wear Volume Loss (mm³/million cycles) Data of Sulzer Metasul® and Experimental As-Cast Co-Cr Hip Components*

Alloy	Sample #	Cup	Head	Total
Sulzer Metasul®	1	5.29	0.256	5.546
(High carbon wrought)	2	4.94	8.896	13.836
	3	1.98	0.145	2.125
	4	1.55	0.163	1.713
Avg. (N=4)		3.44	2.365	5.805
Experimental as-cast	1	1.19	0.48	1.67
	2	3.98	1.76	5.74
Avg. (N=2)		2.59	1.12	3.705

Summary

The wear characteristics of eight different Co-Cr alloy metal-on-metal bearing combinations were studied using a reciprocating wear machine. Wear specimens were made from three different Co-Cr alloys, cast Co-Cr (as-cast and solution treated conditions), low carbon wrought Co-Cr and high carbon wrought Co-Cr alloys. It was found that the as-cast on as-cast couple was superior to the solution treated on solution treated couple or the as-cast on the solution treated couple. Study also showed that as-cast on as-cast couple was superior to the high carbon wrought alloy. The low carbon wrought alloy was found to be the worst alloy for a metal-on-metal wear application.

The as-cast on as-cast head/cup components were also tested using the hip simulator wear machine. Results showed that the as-cast on as-cast couple was comparable if not superior to the new-generation metal-on-metal component which was made of the high carbon wrought Co-Cr alloy.

References

[1] Charnley, J., "Total Prosthetic Replacement of the Hip in Relation to Physiotherapy," *Physiotherapy*, Vol. 54, 1968, pp. 406–411.

[2] Charnley, J., "Low Friction Arthroplasty of the Hip," *Theory and Practice*, Berlin, Springer-Verlag, 1979, pp. 3-15.

[3] Lindholm, R. V. and Puranen, J., "Ring Total Hip Replacement in Osteoarthritis," *Acta Orthop Scand,* Vol.47, 1976, pp. 311-316.

[4] Wilson, J. N. and Scales, J. T., "Loosening of Total Hip Replacements with Cement Fixation," Clinical Findings and Laboratory Studies, *Clin Orthop,* Vol. 72, 1970, pp.145-160.

[5] Bentley, G., Duthie, R. B., "A Comparative Review of the McKee-Farrar and Charnley Total Hip Prostheses," *Clin Orthop*, Vol.95, 1973, pp.127-142.

[6] Djerf, K. and Wahlstrom, O., "Total Hip Replacement Comparison Between McKee-Farrar and Charnley Prostheses in A 5 Year Follow-Up Study. *Arch Orthop Trauma Surg*, Vol. 05, 1986, pp.158-162.

[7] Langenskiold, A. and Paavilainen, T., "Total Replacement of 116 Hips by the McKee-Farrar Prostheses A Preliminary Report," *Clin Orthop* Vol. 95, 1973, pp.143-150

[8] Chapchal, G., and Muller, W., "Total Hip Replacement with the McKee Prosthesis A Study of 121 Follow-up Cases Using Neutral Cement," *Clin Orthop*, Vol. 72, 1970, pp. 115-122.

[9] Schmalzried, T. P., Szuszczewicz, E. D., Akizuki, K. H., Peterson, T. D., and Amstutz, H. C., "Factors Correlating with Long Term Survival of McKee-Farrar Total Hip Prosthesis," *Clin Orthop*, Vol. 329S, 1996, pp.S48-59.

[10] Semlitsch, M., Streicher R. M., and Weber H., "Wear Behavior of Cast CoCrMo Cups and Balls in Long-Term Implanted Total Hip Psostheses," *Orthopaede*, Vol. 18,1989, pp.377-381.

[11] Streicher, R. M., Schon R., and Semlitsch M., "Investigation of the Tribological Behaviour of Metal-on-Metal Combinations for Artificial Hip Joints," *Biomedizinische Technik*, Vol. 35, 1990, pp.3-7.

[12] Paul, J., "Loading on Normal Hip and Knee Joints and on Joint Replacements," *Engineering in Medicine*, Vol. 2, Advances in Artificial Hip and Knee Joint technology, M. Schaldach and D. Hohman (eds.), Springer-Verlag, New York, 1976.

[13] Chan, F. W., Medley, J. B., Krygier, J. J., Podgorsak, G. F. , and Tanzer, M., "Investigation of Parameters Controlling Wear of Metal-Metal Bearings In Total Hip Arthroplasty," *43rd Annual Meeting of Orthopaedic Research Society,* Feb. 9-13, 1997, pp.763.

Kenneth R. St. John,[1] Robert A. Poggie,[2] Lyle D. Zardiackas,[1] and Richard M. Afflitto[2]

Comparison of Two Cobalt-Based Alloys for Use in Metal-on-Metal Hip Prostheses: Evaluation of the Wear Properties in a Simulator

---

Reference: St. John, K. R., Poggie, R. A., Zardiackas, L. D., and Afflitto, R. M., "Comparison of Two Cobalt-Based Alloys for Use in Metal-on-Metal Hip Prostheses: Evaluation of the Wear Properties in a Simulator," *Cobalt-Base Alloys for Biomedical Applications, ASTM STP 1365,* J. A. Disegi, R. L. Kennedy, and R. Pilliar, Eds., American Society for Testing and Materials, West Conshohocken, PA, 1999.

Abstract: One of the areas of research for reducing the generation of wear debris from total joint replacements is the utilization of cobalt alloys that meet ASTM F 1537 for the production of metal on metal bearing pairs. The standard currently allows carbon contents to range between 0% and 0.35%. Some research on metal-on-metal hip bearings indicates that alloys with carbon content at the upper end of this range will yield bearing pairs with superior resistance to wear. The purpose of this study was to compare the wear resistance of components manufactured from high carbon (0.24%) and low carbon (0.06%) alloys when tested in a hip simulator.

Wear of the metal cups was measured by a gravimetric (weight loss) technique every 500 000 cycles to a total of 5 000 000 cycles. Wear of the metal heads was evaluated approximately every 1 500 000 cycles, to 5 000 000 cycles. Results of the study show a small difference in wear between the two alloys. After the initial wear-in period, the samples with the higher carbon content exhibited a significantly lower wear rate than those with the lower carbon content.

Keywords: wear testing, cobalt alloys, hip simulator, carbon composition

---

[1] Assistant Professor and Professor, respectively, Department of Orthopaedic Surgery and Rehabilitation, University of Mississippi Medical Center, 2500 North State Street, Jackson, MS 39216.

[2] Director of Applied Research and Manager, Materials Engineering, respectively, Implex Corporation, 80 Commerce Way, Allendale, NJ 07401.

Introduction

The development of devices for total hip replacement has involved the use of devices that employ metal or ceramic bearing against polyethylene as well as metal bearing against metal. Some of the early metal-on-metal implants did not perform well clinically [1] with reports of loosening and wear debris formation, prompting most manufacturers to work exclusively with systems that incorporate an ultra high molecular weight polyethylene acetabular cup coupled with either a metal or ceramic femoral head component.

As other design problems and failure modes have been eliminated or reduced, it has become clear that, in many patients, the polyethylene acetabular cup may wear excessively, generating particulate debris that has been implicated in osteolysis and device failure in some patients [2]. As particles accumulate in tissues surrounding the device, the cellular response may include the release of biochemical substances that are believed to be responsible for the development of periprosthetic cystic lesions [3] which are a cause for concern in postoperative radiographs and may lead to loss of fixation of the device.

A re-investigation of the long term results of some of the early metal-on-metal hip replacements frequently shows that the results for the devices that did not experience early failure were nearly as good as for metal on polyethylene systems [4]. In recent years, there has been activity toward re-introduction of metal on metal hip prosthesis systems in the United States and it has become clear that, with proper design, many of the problems exhibited by earlier designs are no longer a concern [5,6]. In fact, it has been reported that the original devices of this type that survived without early failures have shown long term (greater than twenty years) success [7]. The alloys currently being used in the manufacture of these components are, almost universally, alloys that meet the requirements of the ASTM Specification for Wrought Cobalt-28 Chromium-6 Molybdenum Alloy for Surgical Implants (F 1537). This document, in its current draft revision, will specify two different types of material depending upon the carbon composition of the alloy. It has been suggested that alloys with higher carbon composition may be superior in wear resistance in prosthetic hip applications due to the size and disposition of the carbides within the alloy, and particularly at the polished bearing surface [8,9].

The purpose of this study was to characterize the wear behavior of the two alloy variants (high and low carbon content) in a hip simulator when bearing against components of the same alloy.

Materials and Methods

Experimental Specimens

The device components evaluated in this study were all supplied by and manufactured by the same facilities (supplier: Implex Corporation, Allendale, NJ; manufacturer: Stratec Medical, Oberdorf, Switzerland) and to the same dimensional specifications. Other than differences in head offset, the only difference between the two groups of samples was the difference in the base alloy. All specimens were manufactured from cobalt-base alloys that conform to ASTM F 1537. The low carbon alloy had a carbon composition of 0.055 – 0.06% (heads: Bohler, address unknown, cups: CCM alloy, CarTech, Reading, PA) while the higher carbon alloy had a carbon composition of 0.24% (CCM+ alloy, CarTech, Reading, PA). Five pairs of components were tested for each of the two alloy compositions. All heads and cups were 28 mm and the low carbon devices were equivalent to those tested by Medley [10]. The specimens had been manufactured and all quality assurance procedures performed to fully qualify them for clinical use (in Europe). The components were supplied and tested non-sterile.

The diameters of the heads and cups were measured at the manufacturer during the quality assurance process. The high carbon pairs had a diametral clearance of 87 μm (one pair), 88 μm (three pairs) or 98 μm (one pair). The low carbon alloy pairs had a diametral clearance of either 85 μm (three pairs) or 86 μm (two pairs). These values are similar to those reported by Medley [10] for his testing and within the range of clearances that are used for these devices clinically in Europe. Surface roughness measurements were not performed on these devices.

Testing Protocol

Testing was carried out on an eight station hip simulator (MTS Systems Corporation, Eden Prairie, MN) with two independent load control channels and load, torque, and displacement transducers mounted on each station. The load curve was based upon the Paul-type curve [11,12] with a maximum load of 3000 N and a rate of 1 Hz. The components were mounted in the anatomic (cup on top) configuration on a rotating block at a 23° angle so that, with rotation of the block, the moving component (the inferior, femoral component) oscillated through a 46° complex movement pattern with each rotation cycle. Lubrication was supplied by bovine calf serum (HyClone, Logan UT) supplemented with 20 mM disodium EDTA dihydrate. While the device components being tested were metallic rather than polymeric, testing was performed in accordance with the ASTM Guide for Gravimetric Wear Assessment of Prosthetic Hip-Designs in Simulator Devices (F 1714). One deviation from the ASTM Guide was the elimination of an anti-bacterial agent from the serum due to safety concerns with sodium azide and concerns about possible chemical reactions between the sodium azide and the cobalt and titanium alloys contained in the hip simulator and device components. Observations made between earlier studies in which sodium azide had been used and subsequent studies without sodium azide suggested that changing the serum every 500 000 cycles was sufficient to prevent bacterial problems in our experience.

Approximately every 500 000 cycles, the test was stopped and all fixturing and device components were cleaned, dried and weighed to a precision of 100 μg. Since weighing the heads separately from the fixtures would require disengaging the taper connection and resultant possible weigh loss due to abrasion between the surfaces, the heads were only removed from the fixtures for weighing about every third time the test was stopped (1 500 000 cycles). Weights were determined for the head plus fixture construct at each cleaning.

Results

As can be seen from Figure 1, the high carbon alloy specimens yielded combined wear results for the head and cup that can be divided into two groups, one with lower wear and one with higher wear. The combined wear results for the low carbon alloy

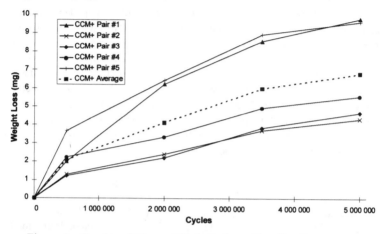

Figure 1 - *Combined Wear of High Carbon Alloy Hip Components*

components (Figure 2) were more reproducible. Investigation of the surface of the two high carbon alloy heads that exhibited higher wear showed a circular defect with a radius nearly identical with that which would be predicted for a scar due to movement against a stationary third body particle in the cup. The scar shown in Figure 3, from one of these heads, has a width of 115 µm. While it was initially speculated that the wear might be

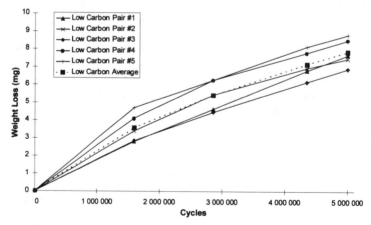

Figure 2 - *Combined Wear of Low Carbon Alloy Components*

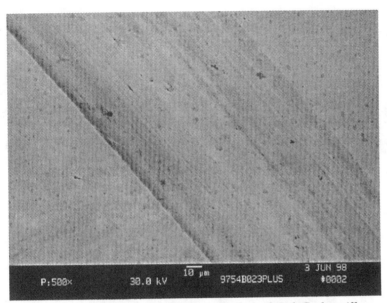

Figure 3 - *Photomicrograph of Scar in Surface of High Carbon Alloy Head with a High Initial Wear Rate*

due to a carbide being removed from one of the bearing surfaces and becoming a third body, this width of scar is inconsistent with the size of the carbides in the high carbon alloy and no likely site of carbide or grain loss was found on either the head or cup component. Analysis of the wear data for the cups (Figure 4) shows that the damage seems to have occurred during the second 500 000 cycles for cup #1 and during the fourth and (possibly) fifth 500 000 cycles for cup #5. Comparison of the average

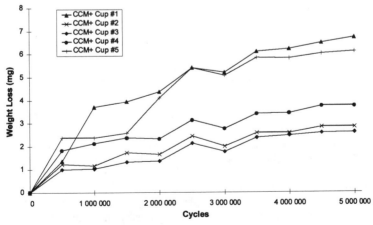

Figure 4 - *Wear of High Carbon Alloy Acetabular Components*

combined wear rates for the two alloys was performed both without (Figure 5) and with (Figure 6) segregation of the higher wear samples from the remaining high carbon alloy samples.

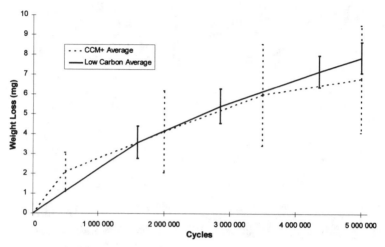

Figure 5 - *Comparison of Average Combined Wear for Both Metal Alloys (Note the High Standard Deviation for the High Carbon Alloy)*

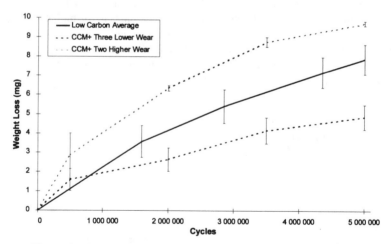

Figure 6 - *Comparison of Average Combine Wear for Both Alloys (Note That the Results for the High Carbon Alloy Have Been Segregated)*

In order to put these metal-on-metal wear rates into perspective, the volumetric wear results for conventional polyethylene samples bearing against cobalt alloy heads are plotted with the metal-on-metal data in Figure 7. The polyethylene wear data was obtained in our laboratory, using the same hip simulator and the same testing protocol.

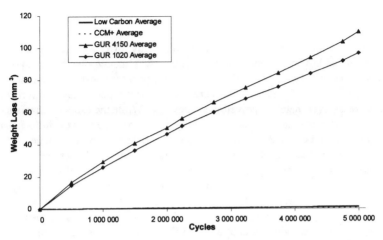

Figure 7 - *Comparison of Average Volumetric Wear for Metal Alloy Systems with Those for Two Different Polyethylene Materials*

Table 1 lists the wear rates for each of the materials in terms of both weight loss and volume of debris generated.

Table 1 - *Wear Rates for Several Hip Bearing Materials Averaged Over Approximately The Last 3 000 000 Cycles (After 2 000 000 Cycles of Run-in)*

Material	Wear Rate mg/million cycles	Wear Rate mm³/million cycles
GUR 4150, Ram Extruded, Sterilized in Vacuum	18.57 ± 1.79	19.75 ± 1.90
GUR 1020, Slab Compression Molded, Sterilized in Vacuum	15.61 ± 0.39	16.60 ± 0.41
Low Carbon Cobalt Alloy (Combined Head and Cup Wear)	1.26 ± 0.10	0.15 ± 0.01
High Carbon Cobalt Alloy (Combined Head and Cup Wear)	0.89 ± 0.23	0.11 ± 0.03
High Carbon Alloy, High Wear	1.13 ± 0.09	0.14 ± 0.01
High Carbon Alloy, Low Wear	0.74 ± 0.09	0.09 ± 0.01

The volume of wear debris was calculated by dividing the weight loss numbers by the reported density of the materials involved. Because of the density differences, the volume of debris corresponding to a particular weight of metal is approximately 12% of the volume that would correspond to the same weight of polyethylene.

Discussion

The results of this experiment show that significantly lower wear rates can be achieved for all-metal hip bearing systems as compared with metal/polyethylene systems with volumetric wear rate ratios of 100:1 to 200:1. This result is consistent with those reported by Chan [13] who found that the wear rates for these types of devices were a 20- to 100-fold decrease over metal on polyethylene devices. While the exact mechanisms of the tissue response to particulate debris are not fully understood, such as the influence of debris size, quantity, and chemistry on tissue response, a reduction in the volume of particulate debris generated and released into surrounding tissues is likely to be beneficial in reducing the potential for osteolysis due to prosthesis wear. Additionally, a reduced volume of wear debris generated translates directly into a reduction in the potentially detrimental changes in the device itself, which may influence device performance.

The higher amount of wear seen for two of the high carbon alloy specimens during one of the run-in periods (initial 500 000 cycles of wear) is similar to the experience of Medley [10], in which one metal-on-metal bearing pair showed a jump in wear during one testing period. Wang [14] also reported one high carbon alloy pair that exhibited high wear compared with other similar components in a similar simulator study. McKellop [7] reported that the retrieved implants that he analyzed also showed evidence of early third body abrasion, which had been polished out over time, with low long term wear. The source of particles that might have caused third body abrasion in the current study has not been determined but the evidence for third body wear occurring is clear. One possible explanation for the observations made in this study and that of Medley is that some event or contaminant causes damage to the metal components causing the release of a few of the small (~2 μm) carbides from the high carbon alloy These carbides then become trapped in the apex of the cup and continue to wear the surface of the moving component while not creating an observable scar on the stationary component. This action would lead to the further release of carbides from the moving component, creating a captive abrasive slurry at the apex of the cup. This slurry could contribute both to wear of the stationary component and widening of the wear scar on the moving component and the slurry would be lost at disassembly for cleaning. This highlights the importance of minimizing the potential for third bodies from the alloy itself and from the surgical procedure, such as bone cement, bone particles, and metal particles due to instrument damage during the procedure.

It is interesting that in the studies by Medley [10], the heads exhibited greater wear than the cups while in this study, the opposite was true. This study was conducted with the components mounted in the anatomic position (cup superior), while their components were mounted in the inverted position. In both studies, the superior (stationary) components experienced higher wear than the inferior (moving) components. Additionally, when apparent third body wear occurred in the high carbon alloy components, the wear scar appeared on the moving component in both studies but the higher level of wear occurred on the stationary component.

Chan [13] evaluated the effect of diametral clearance on the wear rates of metal-on-metal component pairs and concluded that greater clearances yielded higher wear rates. In their study, all of the higher wear specimens were both higher clearance and constructed from low carbon F1537 alloy. In this study, the low carbon pairs had a slightly tighter clearance than the high carbon pairs yet still yielded higher wear rates, suggesting that the alloy composition may have had been a more important factor than component clearance in both studies. While the pair with the highest clearance in our study (98 μm) was one of the two high carbon alloy pairs with higher wear, it seems likely that the third body wear was the primary factor since it exhibited similar wear to the other pair with evidence of third body wear (clearance 88 μm) and both were different from the other two pairs with the 88 μm clearance.

For steady state wear conditions, the wear rate for the higher carbon composition alloy is lower than that for the lower carbon alloy, based both on Figures 5 and 6 and the data in Table 1. Using the data from all samples, the difference in the two wear rates is significant ($p<0.01$) by the Mann-Whitney-Wilcoxon rank sum test. Both metal-on-metal alloys exhibit substantially lower wear rates than metal-on-polyethylene samples tested on the same machine under the same conditions, which is consistent with the work of other researchers. Combining all of the specimens in his study (except the specimen with the third body wear) Medley predicted a total volumetric wear after 24.5 years (24.5 million cycles) of approximately 3.25 mm^3 that translates to 0.13 mm^3/mc, consistent with the results of this study.

Conclusions

The use of metal alloys that conform to the requirements of F1537 to manufacture metal-on-metal hip bearing systems appears to be promising, based upon hip simulator testing. Also, alloys with the higher allowable composition of carbon should provide an advantage in the long term wear resistance, as compared with device components fabricated from lower carbon alloys. Based upon three separate studies that have reported occasional third body damage to high carbon alloy components in hip simulator testing, it is possible that the high carbon alloy may be more susceptible to this type of damage. It is likely that, despite the occurrence of this damage, the long term wear result would be better for the high carbon alloys since even the damaged pairs showed a lower steady state wear rate than the lower carbon alloy.

The volumetric wear rates of the metal-on-metal components tested were at least two orders of magnitude lower than for metal-on-polyethylene components subjected to the same testing conditions. The use of metal bearing surfaces would be expected to be an advantage in achieving long term success of hip prostheses, if wear rate were the only consideration. As mentioned previously, the interaction between debris size and quantity and the effects on tissue must also be considered.

References

[1] Amstutz, H. C., and Grigoris, P., "Metal on Metal Bearings in Hip Arthroplasty," *Clinical Orthopaedics and Related Research,* Vol. 329S, 1996, pp. S11-S34.

[2] Harris, W. H., "Osteolysis and Particle Disease in Hip Replacement: A Review," *Acta Orthopaedica Scandinavica,* Vol. 65, 1994, pp. 113-123.

[3] Jasty, M., Bragdon, C., Jiranek, W., Chandler, H., Maloney, W., and Harris, W. H., "Etiology of Osteolysis Around Porous-Coated Cementless Total Hip Arthroplasties," *Clinical Orthopaedics and Related Research,* Vol. 308, 1994, pp. 111-126.

[4] Jacobsson, S.-A., Djerf, K., and Wahlström, O., "20-Year Results of McKee-Farrar Versus Charnley Prosthesis," *Clinical Orthopaedics and Related Research,* Vol. 329S, 1996, pp. S60-S68.

[5] Müller, M. E., "Lessons of 30 Years of Total Hip Arthroplasty," *Clinical Orthopaedics and Related Research,* Vol. 274, 1992, pp. 12-21.

[6] Weber, B. G., Semlitsch, M. F., and Streicher, R. M., "Total Hip Joint Replacement using a CoCrMo Metal-Metal Sliding Pairing," *Journal of the Japanese Orthopaedic Association,* Vol. 67, 1993, pp. 391-398.

[7] McKellop, H., Park, S.-H., Chiesa, R., Lu, B., Normand, P., Doorn, P., and Amstutz, H., "Twenty Year Wear Analysis of Retrieved Metal-Metal Prostheses," *Transactions of the Fifth World Biomaterials Congress,* II:854, Toronto, Canada, 1996.

[8] Bobyn, J. D., "Alternatives to Metal-on-Polyethylene Bearings," Workshop Speaker, 41st Annual Meeting of the Orthopaedic Research Society, Orlando, Florida, February 15, 1995.

[9] Pilliar, R. M., "Modern Metal Processing for Improved Load Bearing Surgical Implants," *Biomaterials,* Vol. 12, 1991, pp. 95-100.

[10] Medley, J. B., Dowling, J. M., Poggie, R. A., Krygier, J. J., and Bobyn, J. D., "Simulator Wear of Some Commercially Available Metal on Metal Hip Implants," *Alternative Bearing Surfaces in Total Joint Replacement. ASTM STP 1346,* J. J. Jacobs and T. L. Craig, Eds., American Society for Testing and Materials, West Conshohocken, PA, 1998.

[11] Paul, J. P., "Forces Transmitted by Joints in the Human Body," *Proceedings of the Institution of Mechanical Engineers,* Vol. 181(3J), 1967, pp. 8-15.

[12] Paul, J. P., "The Biomechanics of the Hip-joint and its Clinical Relevance," *Proceedings of the Royal Society of Medicine,* Vol. 59, 1966, pp. 943-948.

[13] Chan, F. W., Bobyn, J. D., Medley, J. B., Krygier, J. J., Yue, S., and Tanzer, M., "Wear Performance of Metal-Metal Hip Implants," *Archives of the American Academy of Orthopaedic Surgeons,* Vol. 1, 1997, pp. 57-60.

[14] Wang, K., Wang, A., Gustavson, L., "Metal-on-Metal Wear Testing of Co-Cr Alloys," Symposium on Cobalt-Based Alloys for Biomedical Applications, ASTM, Norfolk, VA, November 3-4, 1998.

John A. Killar,[1] Howard L. Freese,[2] Richard L. Kennedy,[2] and Martine LaBerge[1]

Effect of Metallic Counterpart Selection on the Tribological Properties of UHMWPE

Reference: Killar, J. A., Freese, H. L., Kennedy, R. L., and LaBerge, M., "**Effect of Metallic Counterpart Selection on the Tribological Properties of UHMWPE,**" *Symposium on Cobalt-Base Alloys for Biomedical Applications, ASTM STP 1365*, J.A. Disegi, R. L. Kennedy, and R. Pilliar, Eds., American Society for Testing and Materials, West Conshohocken, PA, 1999.

Abstract: This study was aimed at evaluating the effect of counterpart selection on the wear rate, and the surface and subsurface properties of ultra high molecular weight polyethylene (UHMWPE). The frictional and wear properties of an implant grade UHMWPE in contact with ASTM F 1537 cobalt-based alloy and ASTM F 138 316L stainless steel were characterized. The average contact angle measurements obtained for the 316L SS surfaces were statistically higher (C.I.=95%) than those measured for the Co-Cr-Mo alloy samples. The surface roughness of the metallic counterparts was not significantly altered during wear testing. UHMWPE sub-surface changes as well as wear rate were more pronounced with 316L SS counterparts.

Keywords: UHMWPE, polyethylene, wear, cobalt-base alloy

Introduction

As stated by Jacobs et al. [1], wear has emerged as a central problem limiting the long-term longevity of total joint replacements. Several factors such as cyclic loading, extremely high loads, material processing properties, sterilization, aging, and design attributes have been identified as determinants of the wear behavior of ultra-high molecular weight polyethylene (UHMWPE) [2]. As part of a tribosystem, the wear performance of UHMWPE would be influenced by the selection of the metallic counterpart. High corrosion resistance and superior mechanical properties are primary concerns when selecting

[1]Graduate Assistant and Associate Professor of Bioengineering, respectively, Department of Bioengineering, 501 Rhodes, Clemson University, Clemson, South Carolina, 29634-0905, USA.

[2]Manager, Business Development and Technical Services, and VP of Technology, respectively, Allvac, 2020 Ashcraft Avenue, PO Box 5030, Monroe, North Carolina, 28111-5030, USA.

the counterpart material. Both iron-based alloys and cobalt-chromium-molybdenum (Co-Cr-Mo) alloys have been used as bearing surfaces in total joint replacements. Forged iron-based alloys which show exceptional mechanical properties have replaced cast stainless steels with inadequate yield and fatigue strengths. Iron-based alloys demonstrating high ductility, combined with relatively low cost, have resulted in an increase of interest in the use of 316L stainless steel in joint replacements. Wrought Co-Cr-Mo alloy offers several advantages as an implant material such as increased yield and tensile strengths and fatigue properties, and is largely used as an orthopaedic bearing surface against UHMWPE and against itself as in the metal-metal total hip replacement [3,4]. Wrought Co-Cr-Mo alloy contains less carbon that the cast Co-Cr-Mo alloy (usually 0.05% or lower), and the alloy contains chromium content of 26.0 to 30.0 wt% as compared to 27.0 to 30.0 wt% for cast Co-Cr-Mo alloy [5,6]. Strength and corrosion resistance are improved by adding nitrogen (0.25 wt% maximum). Even though most metallic alloys are largely influenced by excessive localized contact stresses, cobalt-based alloys are more wear resistant than iron based alloys. However, studies have also shown that both 316L stainless steel and Co-Cr-Mo alloy exhibit similar metallic wear resistance against UHMWPE [7]. Critical design parameters for bearing surfaces, excluding the bearing as a whole, include surface roughness and surface hardness uniformity on the bearing surface, among others. Also, the surface chemistry of the surfaces and their wettability directly affect the wear performance of a tribosystem [8].

This study was aimed at testing the hypothesis that the tribological performance of a contact metal-UHMWPE, including friction coefficient, wear rate, and sub-surface changes, is influenced by the selection of the metallic alloy.

Materials and Methods

Ultra-High Molecular Weight Polyethylene

A 5 foot long, 3 inch diameter ram extruded bar of UHMWPE (Poly Hi Solidur from 4150HP UHMWPE powder, lot 854858, 0.93 g/cc density, manufactured from Hoechst Celanese) was obtained from the Hospital for Special Surgery, Department of Biomechanics & Biomaterials (New York, NY). Bars were inspected to conform with the guidelines established by ASTM F 648 method "Ultra-High-Molecular-Weight Polyethylene Powder and Fabricated Form for Surgical Implants," and then annealed by Poly Hi Solidur at a proprietary temperature. Pins of 0.95 cm in diameter by 1.90 cm in length were machined and a lathe was used to achieve a smooth finish on each face as well as along the length. The pins were cleaned [9] and gamma sterilized in air (2.5 Mrads) (Isomedix Operations, Inc., Spartanburg, SC). After sterilization, the UHMWPE pins were weighed, and presoaked (protocol described below) in the test lubricant before testing.

Metallic Alloys

Co-Cr-Mo alloy (ASTM F 1537 method, "Wrought Cobalt-28-Chromium-6-Molybdenum Alloy for Surgical Implants") and 316LSS (ASTM method F 138 "Stainless Steel Bar and Wire for Surgical Implants") were used as counterparts in the form of 2.5 cm wide x 10 cm long x 0.6 cm thick coupons. Each sample was mounted using double stick tape onto a plastic slide for polishing (D-2000 Norderstedt, EXAKT Technologies, Inc., Oklahoma City, OK). This technique involved using paper grit no. 320, 400, 600, 800, 1000, 1200, 2000, alumina powder 5μm and polishing cloth for 30 minutes each at a grinder speed of 100 (instrument scale) and oscillation speed of 80 (instrument scale) and a load of 4 N (excluding sample weight). Each sample was thoroughly washed with soap (Liquinox™) and distilled water between changes in paper grit size. After polishing, the samples were cleaned [9] and dried.

Lubricant

Sodium azide (NaAz, Fisher Scientific: S227-100) and ethylene diamine tetraacetate (EDTA, Gibco Life Technologies: 15576-010) were added (0.2% NaAz, 20mM EDTA) to 100% bovine blood serum (Gibco, by Life Technologies, Gaithersburg, MD) before testing as an antimicrobial agent and calcification inhibitor respectively. The lubricant was changed daily and pooled for filtration and debris analysis.

Test Setup and Conditions

A modified reciprocating friction apparatus (Engineering Machine Shop, University of Waterloo, Ontario) was used to articulate the polyethylene pins against the test metals. The upper bearing surface consisted of a stainless steel jig (Figure 1) which holds three polyethylene pins in a tripod configuration. Three stainless steel (316 SS) collets (Hardinge, Inc., Elmira, NY) were built into the jig and served as grips for the polyethylene pins. The collets prevented toggling or rotation of the pins that could occur due to the articulation. The metals were placed into a specially designed tray and held in place with set screws. A Plexiglass™ lid with three lubricant windows was screwed into the tray and an elastomeric sealant was used to provide a seal between the metal samples and the lid. Fluid loss due to evaporation was replenished by deionized water using a pump (Keofeed II, IVAC Corp, San Diego, CA) at a rate of 3 mL/hr.

The apparatus consists of two parts: 1) a linear motion reciprocating platform driven by a scotch yoke with sinusoidally varying velocity, and 2) force transducers attached to a fixed frame physically isolated from the platform. The displacement of the reciprocating platform from which velocity can be determined is measured by a linear variable displacement transducer (LVDT) (Schaevitz, NJ). The force transducers consist of a cantilever beam of stainless steel (2" long, 0.75" wide, and 0.125" thick) to which four strain gages (350 Ohms) are connected in a full Wheatstone bridge configuration. An approximately 0.2" diameter hole was drilled through each transducer before strain gage mounting and a screw and nut were used to attach the force transducer to the specimen jig. The output from the strain gages was amplified, collected, and calibrated using a 12 bit

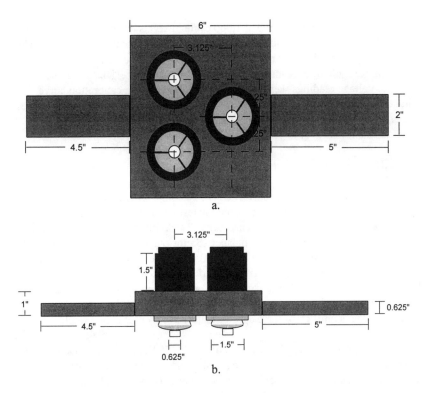

Figure1 - *Specimen jig holding polyethylene pins. a.) Bottom view b.) Side view.*

analog-to-digital converter (10V range, National Instruments) and a personal Macintosh IIsi computer with a data analysis software package (LabView II, National Instruments). The temperature of the lubricant was maintained at approximately 37°C during testing using a rectangular silicon rubber heater (SRFG-509, Omega Engineering, Inc.) placed under the counterface specimen holder and a copper/constantan thermocouple (0.01" diameter, PL016019, Omega Engineering, Inc.) driven by a temperature controller through a solid state relay.

Test Conditions

A nominal contact stress of 3.45 MPa was achieved using an axial load of 735 N (245 N for each pin). Specimens were tested for 1.5×10^6 cycles for a total sliding distance of 150 km at a frequency of 1 cycle/sec and a sliding distance of 100 mm/cycle. Wear data were measured by weight loss after 50 km (500,000 cycles), 100 km (1×10^6 cycles), and 150 km (1.5×10^6 cycles). For each weighing, the test apparatus was stopped, unloaded, force transducers disengaged, and the UHMWPE pins removed. The pins were cleaned, dried, weighed, and replaced into the jig collets.

Friction measurements were acquired at 1,000 cycle intervals during sliding. Friction data were collected every 0.04 seconds for 10 seconds (250 data points per reading). The corresponding voltage recordings from the strain gages on the force transducers were stored in a format readable by spreadsheet software (Microsoft Excel®). From the spreadsheet, the voltage readings were converted to a force using a calibration curve. LVDT voltage readings were also taken at each cycle and sliding distances and velocities could be determined using a process similar to that of the force transducers with a calibration curve for the LVDT. A total of six couples were tested per group.

Pre-soaking and Creep Measurement

The UHMWPE pins were positioned vertically in a container and lubricant was added to approximate the level of fluid which occupied the test windows (≈0.6 cm). Presoaked specimens were kept in an environmental chamber at 37°C. Fluid absorption was measured by weight change using an analytical balance (Sartorious Corp, Bohemia, NY) accurate to ±0.00002g. Before weighing, the UHMWPE pins were cleaned (ASTM method F 732-91 "Reciprocating Pin-on-Flat Evaluation of Friction and Wear Properties of Polymeric Materials for Use in Total Joint Prostheses" (not reapproved in 1997)). Each pin was weighed four times and weights were averaged to minimize random error. Pins were weighed every two days until gains were approximately 20μg or less, then weekly until testing began.

A creep test was performed applying a load of 735 N (75 kg) (wear test load per pin) and for a duration equivalent to that of the wear test. Loaded pre-soaked control pins were cleaned and weighed using the protocol mentioned above. Dimensional measurements were taken of each of the control pins for creep measurement and averaged before the load was applied. The pins were set-up in a triangle configuration as used for wear testing. At time intervals corresponding to each 50 km stop of the wear test, the loaded control pins were unloaded, cleaned, and weighed. Dimensions were measured before applying the load. The weight changed was used to generate net wear data. One pin of each presoaked and creep group was embedded and sectioned according to the aforementioned procedure for an optic microscopy analysis.

At the completion of wear tests, the corrected wear rate, volume loss, height loss, and wear factor were determined for each pin.

Surface Characterization

Non-contact surface profilometry (TOPO-3D, Wyko Corporation, Tucson, AZ with a 20X Mirau magnification head and a cut-off area of 500 x 500 μm) and scanning electron microscopy (SEM, Jeol-JSM-IC848) were conducted on the metal specimens before testing and along the wear track after testing. The surface of the UHMWPE pins was analyzed using non-contact surface profilometry (NT-2000, Wyko Corporation, Tucson, AZ , 5.1X) and atomic force microscopy (AFM, Discoverer SPM, TopoMetrix, Inc., Santa Barbara, CA) before, during, and after testing. Five measurements were taken and average roughness (RA), root mean square (RMS) roughness, and peak-to-valley (P-V) ranges were calculated for each specimen. At the completion of the tests, the UHMWPE pins were gold coated (Hummer X, Anatech LTD, Alexandria, VA) and scanning electron microscopy analysis was performed on the bearing surface.

Critical Surface Tension and Subsurface Characterization

Contact angle measurements were obtained using a goniometer (Gaertner Scientific Corp., Chicago, IL). Four to five drops of three test fluids (water, HPLC grade (Sigma, 27,073-3), methylene iodide (Fisher, M-227), and hexadecane (Fisher, 03035)) of known liquid/vapor surface tension were placed onto each sample and the advancing and receeding contact angles were measured. The critical surface tension of the test material was then calculated [*10*]. This technique did not follow any ASTM methods.

Test UHMWPE specimens as well as one loaded control pin and one non-loaded control pin were embedded in glycol methacrylate (Technovit 7100, Kulzer, Germany). Longitudinal sections of 10 μm thick were cut using a tungsten carbide blade (Polycut E, Reichert-Jung, Buffalo, NY) to examine subsurface failure and deformation using polarized microscopy.

Statistical Analysis

A statistical analysis using a general linear model with variance analyses ($p \leq 0.05$) was performed on all roughness, creep, friction, and wear measurements. Surface roughness, creep, and wear measurements were analyzed using Microsoft Excel® Data Analysis Tools. A separate program (general linear model) was written using Statistical Analysis Software (SAS Corp, Cary, NC) to compare the friction measurements for each test at each cycle reading.

Results

SEM and profilometry showed that a mirror surface finish was achieved on all metal samples. Average RMS, Ra, and PV values of 5.01 ± 0.94 nm, 4.05 ± 0.76 nm, and 37.13 ± 3.85 nm respectively were measured for the Co-Cr-Mo alloy specimens, and 5.02 ± 0.15, 3.97 ± 0.13, 40.95 ± 8.79 nm for the 316L SS specimens. Roughness parameters for both Co-Cr-Mo alloy and 316L SS were not significantly different (ANOVA, $p \geq 0.05$). An average 2% increase in roughness measurements for the Co-Cr-Mo alloy specimens (5.18 ± 1.08 nm) and a decrease of 15% for the 316L SS specimens (4.28 ± 0.49 nm) were observed after testing. However, these differences were not significantly different ($p \geq 0.05$).

A pattern of concentric circles produced by the lathe during pin fabrication was observed using SEM. The center of the pin has a rougher appearance than the outer fabrication grooves. This is supported by the roughness measurements calculated from the atomic force images. The AFM scan performed near the center of the pin showed an av-

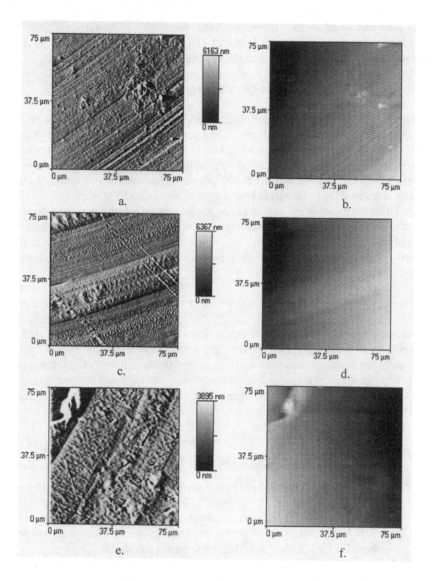

Figure 2 - *Representative AFM images (force mode (a,c,e) and corresponding height mode (b,d,f)) of unworn control UHMWPE. a.-b.) area near the center, c.-d.) area midway between the center and edge, and. e.-f.) near the outer edge of the pin.*

erage RMS value of 2.06 µm, while RMS values of 0.77 to 1.148 µm were obtained for peripheral regions (Figure 2). Similar observations were obtained with the non-contacting profilometer. After wear testing, the concentric machine grooves were flattened or totally worn away and new scratches and troughs were evident in the direction of the articulation (Figure 3). Average RMS roughness of 1.48 ± 0.64 µm was measured for the UHMWPE samples before testing. After testing, an average RMS roughness of 1.44 ± 0.44 µm and 0.94 ± 0.66 µm was determined for the Co-Cr-Mo alloy group and the 316L SS group, respectively. RMS roughness measurements were not significantly different (ANOVA, $p \geq 0.05$).

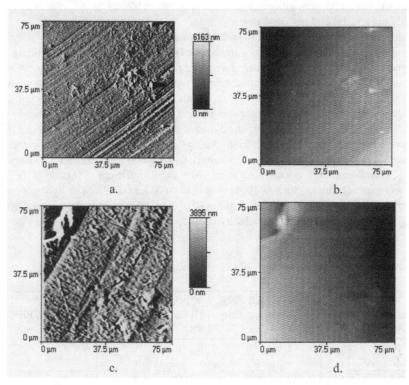

Figure 3 - *Representative AFM images (force mode (a,c) and corresponding height mode (b,d)) of worn UHMWPE pins. a.-b.) pin worn against Co-Cr-Mo alloy, c.-d.) pin worn against 316L SS.*

A critical surface tension (CST) of 31.6 dynes/cm was measured for UHMWPE. Polished orthopaedic metals generally have a CST greater than 40 dynes/cm; however, the CST measured in this study were approximately 30 dynes/cm for both metals [10]. This difference may be attributed to the lower surface roughness of the metals used in this study. The RMS roughness of orthopaedic implants are approximately 10 times greater than the RMS roughness of the test alloys in this study. A roughened surface will have a higher CST than a smooth surface of the same material [11]. A contact angle of 58.17 ± 2.14 for the Co-Cr-Mo alloy specimens measured with HPLC grade was found to be significantly lower than the contact angle measured for the 316L SS specimens (74.88 ± 0.83) (ANOVA, $p < 0.05$).

Wear calculations were performed using weight loss data obtained during steady state articulation. Even though lower, the wear rate of UHMWPE pins articulating against Co-Cr-Mo was not statistically different than that of the 316L SS group ($p \geq 0.05$) (Table 1). Results summarized in Table 1 are comparable to similar studies in the literature although, direct comparison with other research is questionable due to differences in materials (manufacturer) and test conditions.

The static coefficient of friction (μ) was determined at start-up. The low velocity and high velocity coefficients of friction were extrapolated from the data set for a sliding velocity equal to or less than 30 mm/sec and greater than 30 mm/sec, respectively. The static μ remained fairly constant from the initial test start through both test restarts with an average of 0.03 for both the Co-Cr-Mo alloy and 316L SS groups. At the test onset and after each restart, a gradual decrease in friction occurred with time until a steady state value was reached. At low velocity, the dynamic μ for both Co-Cr-Mo alloy and 316L SS groups were not significantly different during the overall testing period and were comparable to the static coefficient of friction (0.014 ± 0.006 and 0.012 ± 0.005 respectively). However at high velocity, the average dynamic coefficient of friction of the 316L SS group increased by as much as 100% over time (0.024 ± 0.002). The dynamic coefficient of friction of the Co-Cr-Mo alloy group (0.016 ± 0.005) remained constant during testing and comparable to the low velocity coefficient of friction.

Table I. *Wear parameters of Poly Hi Solidur pins vs. test counterparts.*

Average Steady State Wear Calculations	Experimental Poly Hi Solidur vs. Co-Cr-Mo Alloy	Experimental Poly Hi Solidur vs. 316L SS	Wear data by others for UHMWPE/Co-Cr alloy contact
Calculated Wear Rate (mg/10^6)	0.117 ± 0.010	0.180 ± 0.104	Not Reported
Corrected* Wear Rate (mg/10^6)	0.130 ± 0.010	0.199 ± 0.104	0.18mg/10^6 [12]
Volume Loss (ρ=0.93, mm³/10^6)	0.139 ± 0.011	0.152 ± 0.032	0.188 mm³/10^6 [12]
Height Loss (μm/10^6)	1.958 ± 0.157	2.127 ± 0.454	2.68μm/10^6 [12]
Wear Factor ((mm³/(N•m)*10^{-7})	0.799 ± 0.064	0.868 ± 0.185	1.810 (for SS) mm³/(Nm)*10^{-7}[13]

*Weight changes measured in loaded control tests were used to correct for creep during testing according to Dowson et al. [14].

Under polarized light microscopy, the control specimens showed an imperfect consolidation of the UHMWPE grains. Grain boundaries and unfused grains were apparent. Also, the machined surface showed surface waviness, roughness and a narrow band of polarized light at a different wavelength. After wear testing, the thickness of this band in test specimens was greater than for the control specimens. An image analysis software (NIH 2.1) used to quantify the thickness change and the level of consolidation of the grains within and around that band, demonstrated that the thickness of the band was larger by a factor of 25% in the 316L SS group. Also, the asperities of the worn face were flattened or abraded which produced a smoother surface appearance than observed on the control pins. Cracks were also evident at the base of some asperities on the worn pin faces which would suggest that fatigue or delamination wear occurred during testing (Figure 4).

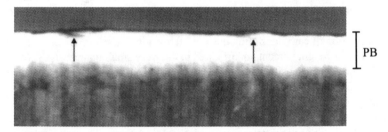

Figure 4 - *Representative UHMWPE samples (316L SS group)*
after wear testing showing cracks (see arrows). PB (Polarized light band).

Discussion

Co-Cr alloys are commonly used in the design of orthopaedic artificial joints because of their high corrosion resistance and their mechanical properties that provide reliability and performance in highly demanding dynamic conditions. Several tribological studies have been conducted to evaluate the tribological properties of these alloys using UHMWPE as a counterpart. More recently, the object of these studies was the characterization of the wear performance of UHMWPE as UHMWPE debris have been shown to be the leading cause of osteolysis and implant loosening [1].

This study compared the tribological properties of two orthopaedic metals that are suitable for bearing application. The results have demonstrated that the selection of the metallic counterpart can affect the overall tribological properties of UHMWPE. Even though the wear rate of UHMWPE was not influenced by the metallic counterpart, the frictional behavior of the UHMWPE-Co-Cr-Mo alloy tribosystem was more desirable than that observed with the UHMWPE-316L SS tribosystem. The friction force observed on the surface is translated in terms of shear stress. One consequence of the variable friction observed in this experiment is a stress concentration. An analytical model by Joseph et al. [15] shows that a high gradient in the coefficient of friction will cause a strain localization. Also, variable friction can be translated in terms of fatigue that will contribute to enhance the surface change of the UHMWPE and contribute to crack initiation and growth. Therefore, it is realistically assumed that uniform friction properties minimize surface damage. Since variable friction was observed with the 316L SS, using frictional forces as a unique contribution to surface damage, it could be hypothesized that the wear rate of UHMWPE articulating against 316L SS would have been higher than with Co-Cr-Mo alloy as a counterpart for a reciprocating distance longer than 150 km.

Despite its low and constant coefficient of friction as well as a contact angle that favors boundary lubrication, Co-Cr-Mo alloy contributed to damage the surface of the UHMWPE counterparts. Surface analysis and electron microscopy indicated that the surface change of UHMWPE articulating against Co-Cr-Mo alloy and 316L SS were roughening and smoothening respectively even though the surface roughness of both metals were similar before and after testing. Even though several measurements were taken on the same surface to minimize the effect of small cut-off area and magnification inherent to the non-contact surface profilometry technique, large asperities might have been overlooked. At a microscopic level, the presence of carbon in the cobalt based alloys will result in the formation of carbides during the alloy fabrication. Carbon content can vary from 0% to 0.35% weight percent and is added in this very small quantity (ASTM method F 1537 alloy composition is used for forged medical and surgical components, and ASTM F 75 alloy composition is used for castings) to form carbides which serve as a strengthening mechanism for cobalt based alloys. Carbides that precipitate at the grain boundaries and within the grains prevent grain boundary sliding which contributes to strengthening [16]. The precipitation of carbides is not uniform. The resulting surface presents regions of different hardness which, consequently, may affect the wear behavior of the tribosystem when used against a polymer such as UHMWPE.

SEM, AFM, and optical observations suggest that the mechanisms which caused wear of the polyethylene surface articulating against the metals included mainly abrasion and delamination. The flat-on-flat contact geometry, as used in this study, does not promote full fluid lubrication. Therefore, direct or partial contact between the UHMWPE asperities and the metal asperities have occurred depending on the level of surface protection due to boundary lubrication. A transfer film was not evident on the metal samples after testing. Delamination is normally observed when a subsurface deformation forms usually due to stress or creep which causes the initiation of a crack which then grows and leads to the detachment of a layer of polyethylene from the surface. In this respect, the presence of the light colored bands just below the bearing surface indicate that the material observed some residual strains during loading and sliding which could provide an initiation point for delamination. The most probable location for delamination to occur is at the edges of the pin where the contact stress is highest and as demonstrated through polarized microscopy. UHMWPE sub-surface changes were more apparent with 316L SS as a counterpart material.

Standard lubrication theory requires the generation of a fluid wedge that acts to separate articulating surfaces and minimize friction [17]. The theory applies to contact geometry's such as point contacts and line contacts which are conducive to wedge formation. Flat-on-flat contacts are not leading to wedge generation. Since the friction measured in this experiment (average high velocity $\mu \approx 0.013$) was lower than friction reported in non-lubricated contacts ($\mu = 0.13$, Semlitsch et al.[18]) and lubricated contacts, it is possible that some surface separation by the lubricant occurred. The bovine serum contains several proteins, which may be adhering to the articulating surfaces and functioning as a boundary lubricant which could in part explain a higher friction immediately after a lubricant change which would disrupt protein adherence.

The test set-up used to conduct this wear study provided a linear reciprocating motion of the contact. Recently, a study comparing the wear behavior of the metal-UHMWPE contact with systems using multidirectional (or "cross-shear") motion and linear motion, has shown that wear rates are dependent on the direction of the motion [19]. Therefore, this study on the effect of alloy selection of UHMWPE wear should be repeated with a multidirectional motion.

Conclusion

Co-Cr-Mo alloy is widely used in total joint replacement designs primarily for its high fatigue strength, corrosion resistance, and ease of manufacture. This study showed that it is also a desirable counterpart as compared to 316L SS when sliding against UHMWPE. The wear behavior of UHMWPE can be affected by the selection of the metallic counterpart. UHMWPE sub-surface changes were more pronounced with 316L SS counterparts than Co-Cr-Mo alloy counterparts.

Acknowledgements

The authors thank Dr. Stephen Li of the Hospital for Special Surgery (NY) for providing the UHMWPE. Dr. Michael J. Drews and Ms. Kim Ivey of Clemson University are acknowledged for providing expertise for the polymer chemical analysis. This work was allowed through the support of Allvac, Monroe, NC, and The National Science Foundation grant number CMS-9601859.

References

[1] Jacobs, J.J., Shanbhag, A., Glant, T.T., Black, J., and Galante J.O., "Wear debris in total joint replacements," *Journal of the American Academy of Orthopaedic Surgery.*, 1994, vol. 2, pp. 212-220.
[2] AAOS, *Implant Wear: The Future of Total Joint Replacement*, Wright and Goodman Ed., 1996.
[3] Medley, .JB., Krygier, J.J., Bobyn, J.D., Chan, F.W., and Tanzer, M., "Metal-metal bearing surfaces in the hip: Investigation of factors influencing wear," *Transactions of the Orthopaedic Research Society*, 1995, vol. 20(2), pp. 765.
[4] Streicher, R.M., Schon, R., and Semlitsch, M., "Investigation of the tribological behaviour of metal-on-metal combinations for artificial hip joints," *Biomed. Tech.*, 1990, vol. 35(5),pp. 3-7.
[5] Gomez, M., Mancha, H., Salinas, A., Rodriguez, J.L., Escobedo, J., and Castro, M., Mendez M., "Relationship between microstructure and ductility of investment

cast ASTM F-75 implant alloy," *Journal of Biomedical Materials Research*, 1997, vol. 34, pp. 157-163.

[6] Shetty, R.H., and Ottersberg, W.H., "Metals in orthopaedic surgery," in *Encyclopedic Handbook of Biomaterials and Bioengineering*, Wise, Trantolo, Altobelli, Yaszemski, Gresser, and Schwartz eds. Marcel Dekker, Inc., New York, Part B: Applications, 1995, Vol. 1, 1995, pp. 509-540.

[7] Mittlmeier, T., Walter, A., "The influence of prosthesis design on wear and loosening phenomena," *CRC Critical Reviews in Biocompatibility*, 1987, vol. 3, pp. 319-419.

[8] Rogers, J.M.. Powell, G.L., Pace, T., and LaBerge, M., "Effect of phospholipidic boundary lubrication in rigid and compliant hemiarthroplasty model," In Press, *Journal of Engineering in Medicine*, 1998.

[9] Rowland, S.A., Shalaby, S.W., Latour, R.A., and vonRecum, A.F., "Effectiveness of cleaning surgical implants: Quantitative analysis of contaminant removal," *Journal of Applied Biomaterials*, 1995, vol. 6, pp. 1-8.

[10] Zisman, W., "Relation of equilibirum contact angle to liquid and solid consitution," In Contact Angle, Wettability, and Adhsesion. Advances in Chemistry Series 43, Fowkes, Ed., ACS Press, Washington, 1964, pp. 1-51.

[11] Baier R.E., Meyer A.E., "Surface Analysis," Chapter 8, *Handbook of Biomaterials Evaluation*, von Recum, Ed., Macmillan Publishing Company, New York, 1986, pp. 97-108.

[12] McKellop, H.A., Clarke, I.C., Markolf, K.L., and Amstutz, H.C., "Friction and wear properties of polymer, metal, and ceramic prosthetic joint materials evaluated on a multichannel screening device," *Journal of Biomedical Materials Research*, 1981, vol. 15, pp. 619-653.

[13] Kumar, P., Oka, M., Ikeuchi, K., Shimizu, K., Yamamuro, T., Okumura, H., and Kotoura, Y., "Low wear rate of UHMWPE against zirconia ceramic (Y-PSZ) in comparison to alumina ceramic and SUS 316L alloy," *Journal of Biomedical Materials Research*, 1991, vol. 25, pp. 813-828.

[14] Dowson, D., "A comparative study of the performance of metallic and ceramic femoral head components in total replacement hip joints," *Wear*, 1995, vol. 190, pp. 171-183.

[15] Joseph, P.F., Zhang, N., Gadi, K.S., Flood, L.M., and Kaya, A.C., "Separable and Non-separable Solutions in Singular Stress Analysis," *Transactions of the Joint 1997 ASME/ASCE/SES Summer Meeting*, 1997, p. 704.

[16] Darby, Jr., J.B., and Beck, P.A., "Intermetallic phases in the Co-Cr-Mo system at 1300°C," *Journal of Metals, Transactions AIME*, 1955, June, pp. 765-766.

[17] Fisher, J., and Dowson, D., "Tribology of total artificial joints," *Proceedings of the Institution of Mechanical Engineers,* 1991, vol. 205, pp. 73-79.

[18] Semlitsch, M., Lehmann, M., Weber, H., Doerre, E., and Willert, H.G., "New prospects for a prolonged functional life-span of artificial hip joints by using the material combination polyethylene/aluminium oxide ceramic/metal," *Journal of Biomedical Materials Research*, 1977, vol. 11, pp. 537-552.

[19] Saikko, V., "A multidirectional motion pin-on-disc wear test method for prosthetic materials," *Journal of Biomedical Materials Research*, 1998, vol. 41, pp.58-64.

Mark A. Pellman [1]

An Overview of PVD Coating Development for Co-Based Alloys

Reference: Pellman, M.A., "**An Overview of PVD Coating Development for Co-based Alloys,**" *Cobalt-Base Alloys for Biomedical Applications, ASTM STP 1365,* J.A. Disegi, R.L. Kennedy, R. Pilliar, Eds., American Society for Testing and Materials, West Conshohocken, PA, 1999.

Abstract: PVD coatings such as titanium nitride (TiN), zirconium nitride (ZrN), and more recently diamond like carbon (DLC), can improve the corrosion and abrasion resistance of medical devices. The success of these coatings has been based on their ability to improve wear resistance and corrosion resistance. The materials properties, biocompatibility and performance of commercially available PVD coatings are discussed along with applications for orthopedic implants and dental devices. The next generation of coatings for biomaterials applications is in development. These coatings include tetrahedral amorphous carbon and DLC coatings. The differences between these processes and traditional PVD will be reviewed. In addition, preliminary performance and tribological data will be presented.

Keywords: Co-base alloys, PVD coatings, titanium nitride, DLC, tetrahedral amorphous carbon

Introduction

Since its first introduction to the medical device industry in the late 1980s, Physical Vapor Deposition (PVD) has become widely used to deposit wear resistant thin film coatings on a variety of medical devices including orthopedic implants, pacemakers, surgical instruments, orthodontic appliances and dental instruments. The value of PVD technology rests in its ability to modify the surface properties of a device without changing the underlying materials properties and bio-mechanical functionality [1,2,3,4].

---

[1] Director of Marketing, Multi-Arc Inc., 200 Roundhill Dr., Rockaway, NJ 07866

PVD Coating Properties

A variety of PVD coatings have been shown to be biocompatible including titanium nitride (TiN), zirconium nitride (ZrN), chromium nitride (CrN), titanium aluminum nitride (TiAlN), aluminum titanium nitride (AlTiN), amorphous diamond-like carbon (DLC) and tetrahedral amorphous carbon. Biocompatibility tests were conducted in accordance with ISO 10993-1 guidelines for materials which experience short term body contact. The results (Table 1) indicate that these coatings are acceptable for external and internal medical devices that come into contact with bone, skin, tissue or blood.

Table 1 - *Short Term Biocompatibility Test Results*[1,2]

TEST	RESULT
Sensitization	No evidence of causing delayed dermal contact sensitization in a guinea pig.
Cytotoxicity	No evidence of causing cell lysis or toxicity.
Acute Systemic Toxicity, T12	No systemic toxicity or mortality.
Intracutaneous Toxicity, T13	No evidence of significant irritation or toxicity in rabbits.
Genotoxicity	Not mutagenic
USP Muscle Implantation, T14	Non-irritating to muscle tissue.
Hemolysis	Nonhemolytic, compatible with blood.

[1] Tests performed per ISO 10993-1 at NAMSA.
[2] TiN, ZrN, CrN, TiAlN, AlTiN, DLC and tetrahedral amorphous carbon coatings.

The hardness of PVD coatings range from 2500 HV for TiN to over 8000 HV for tetrahedral amorphous carbon. In addition, the friction between metal components is reduced, particularly with diamond like carbon (DLC) and tetrahedral amorphous carbon films. PVD coatings also provide protection against corrosion [5], oxidation [6,7] and metal ion release from medical devices.

Tetrahedral amorphous carbon and DLC are relatively new classes of coating with significantly different properties despite the similarity of their names. They illustrate the range of structures and properties now possible via the plethora of PVD and plasma assisted Chemical Vapor Deposition (CVD) processes which have been developed over the last decade.

Tetrahedral amorphous carbon, which is deposited by filtered and enhanced arc processes, contains a high proportion of sp^3 bonded carbon and no hydrogen. This gives this class of coatings exceptionally high hardness and higher thermal stability than DLC films (Table 2). DLC films are hydrogenated and may contain other alloying elements including silicon (Si), titanium (Ti) and tungsten (W). Most commercial DLC films are either hydrogenated DLC or metal containing hydrogenated DLC. Coating properties vary significantly depending on the hydrogen and metal content of the film [8,9,10]. In addition, interlayers, multilayer structures and process conditions can be used to influence coating properties.

Table 2 - *Amorphous Carbon Coating Properties*

	Tetrahedral Amorphous Carbon[1]	Hydrogenated DLC[2]	Me-C:H [3]
Hardness (Hv)	8000	4000-6000	600-1200
Thermal Stability (°C)	>500	250-300	250-350
Bonding	primarily sp^3	sp^3 and sp^2	sp^3 and sp^2
Composition	100% Carbon	Carbon + up to 50% H	Carbon + 5-15% Ti/W + up to 50% H

[1] Deposited by enhanced arc PVD.
[2] Hydrogenated DLC deposited by Radio Frequency (RF) plasma CVD.
[3] Metal containing hydrogenated DLC deposited by unbalanced magnetron sputtering.

Wear and Corrosion Test Results

TiN coating has been in clinical use on a variety of orthopedic implants since the early 1990s in North America as well as Europe [11, 12, 13, 14, 15, 16, 17]. The most common applications involve titanium alloy, articulating and non-articulating components, for hip, knee, shoulder and ankle implants. Implant components manufactured from Co-based alloys are not commercially coated; however, partial dentures and false teeth cast from Co-based alloys have been coated with TiN for many years in countries of the former Soviet Union.

Researchers have reported that TiN reduces ion release levels from co-based alloys as well as titanium and stainless steels [18, 19, 20, 21]. In addition, a number of researchers have found TiN to be biocompatible [22, 23, 24, 25, 26, 27].

Behrndt et al [28] have reported preclinical and clinical results for TiN coated Co-Cr-Mo denture castings. Their investigations were performed to determine the oral compatibility of TiN. They tested the abrasion resistance of uncoated and coated castings with a rotating disk probe. The load was held constant at 1 N and the rotational speed was 400 revolutions per minute. The TiN coating reduced the abrasion of Co-Cr-Mo probes by a factor of 4.8 compared to uncoated probes. This was consistent with their clinical longevity test.

Their clinical study included 100 cast dentures. Dentures were examined for wear after 1.7 and 2.5 years. Independent of the length of time in-vivo, the dentures showed no evidence of spalling or abrasion in areas where masticatory stresses existed. Microscopic examination of dentures revealed uncoated areas associated with casting porosity; however, no enlargement of these spots was observed over the course of the wear period.

Knotek, Loffler and Weitkamp [29] also have examined the performance of PVD TiN coated Co-based alloy dentures. They found that TiN is particularly suitable for this application, combining great surface hardness with known biocompatibility. Cast dentures were coated with single layer and two-layer TiN using the cathodic arc process and tested for pitting corrosion resistance. Substantial corrosion was observed on uncoated dentures after only 100 hours immersion in a physiological saline solution containing 0.9% NaCl at 60 °C. Dentures with single-layer coatings exhibited pitting corrosion at a few points on the coated casting surface. No corrosion was visually observed on dentures with a two-layer TiN coating.

The pitting corrosion resistance of Co-based alloys have also been studied by Wisbey, Gregson and Tuke [30]. They reported that the pitting corrosion resistance of Co-Cr-Mo alloy is enhanced by the application of a TiN coating. This was accompanied by a dramatic reduction in the release of metal ions from the Co-Cr-Mo substrate.

Co-Cr-Mo alloy conforming to British standard BS 3531 was investment cast, solution annealed and hot isostatically pressed at 1195 °C. Corrosion test samples were machined from femoral hip components, then coated with TiN via electron beam evaporation. Electrochemical tests were carried out on coated and uncoated specimens in a solution of 0.17 mol NaCl at a PH of 6.2. To investigate the effect of surface finish, polished and grit blasted samples were tested.

Results showed that the pitting potential was increased on both polished and grit blasted samples but the TiN was most effective in raising the pitting potential on the polished surface (Table 3). The pitting potential was increased on this sample from 0.85 to 1.11 V (SCE).

Table 3 - *Pitting Potentials of Co-Cr-Mo in Nonaerated,*
Buffered 0.17 Mol NaCL Solution*

Surface Condition	Uncoated		Coated	
	Polished	Grit blasted	Polished	Grit blasted
Pitting Potential (V)	0.85	0.84	1.11	0.88

* From Wisby et al.

Wisby et al also measured metal ion release from coated and uncoated Co-Cr-Mo samples in a solution of 0.17 mol NaCl and 2.7 x 10^{-3} mol EDTA maintained at 37 °C. The total concentrations of ions released into solution after 550 hours were determined using atomic absorption (Figure 1).

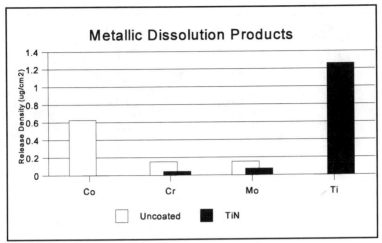

Figure 1 - *Per Wisbey et al [30]*

One of the most critical properties with respect to thin film coatings is adhesion. On medical devices, consistent, high adhesion between the coating and substrate material is required to assure that third body wear debris is not created by coating failure when subjected to loads invivo.

Takeuchi et al [31] measured the adhesion of cathodic arc PVD coated TiN on various orthopedic alloys using a scratch tester. They reported that the TiN coatings exhibited good adhesion on all the substrates tested, particularly on Co-Cr alloy (Table 4).

Adhesion was measured by scratching the surface of coated samples with a diamond stylus at a constant speed and increasing load. Acoustic emission measurements were used to determine the initial load at which the coating began to fail (lower critical load or LCL). Tangential force measurements were used to determine the load at which the coating was scratched from the surface (Upper Critical Load or UCL). Upper Critical Load readings above 45 are generally accepted as an indication of good adhesion. Measured UCL's ranged from 44.7N. for unpassivated Ti-6-4 to 93.3N. for unpassivated Co-Cr-Mo.

Articulating components are susceptible to abrasion by third body particles of metallic wear debris, bone or bone cement. In the case of commercially available total joint replacements, these particles are trapped between the metal (Co-Cr-Mo or Ti-6-4) polyethylene (UHMWPE) components. The abrasive particles can cause damage to both surfaces. Scratches to the metal component, however, are most detrimental to the long term viability of an implant because they accelerate the wear of polyethylene. The polyethylene wear debris then causes osteolysis and bone resorption around the implant which leads to loosening, pain and eventually failure.

Table 4 - *Scratch Test Results for TiN Coated Materials*[1]

Material	Condition[2]	Roughness, (µm)	LCL (N)	UCL (N)
316L	P	0.053	24.7	46.7
316L	NP	0.031	26.7	50.0
17-4 PH	P	0.032	24.0	50.7
17-4 PH	NP	0.039	28.3	49.0
17-4 PH	NP, H900	0.02-0.03	19.8	67.1
Ti-6-4	P	0.165	28.3	50.0
Ti-6-4	NP	0.180	30.3	44.7
Co-Cr-Mo	P	0.033	44.3	90.0
Co-Cr-Mo	NP	0.054	50.7	93.3

[1] Takeuchi et al [31].
[2] P = passivated, NP = not passivated

Poggie et al [32] have tested the abrasion resistance of a tetrahedral amorphous carbon (*TETRABOND*®)[2] and a hydrogenated DLC on Co-Cr alloy in a reciprocating pin-on-disc abrasion test (Table 5).

Table 5 - *Abrasion Performance*[1]

Coating	@1 Million Cycle	@ 5 Million Cycles
TiN	Measurable wear	Worn Through
DLC	Worn through.	Worn Through
Tetrahedral Amorphous Carbon	No visible wear	No Visible Wear

[1] Per Poggie, R [32].

The coatings were subjected to reciprocating sliding abrasion in a pin-on-disc apparatus. Testing was done in Ringer's solution at 2.5 Hz against PMMA bone cement pins. Uncoated samples suffered severe abrasion within very few cycles. TiN coated Co-Cr-Mo pins showed measurable wear at 1 million cycles and wore through prior to 5

[2]*TETRABOND*® is a registered trademark of Multi-Arc Inc., Rockaway, NJ.

million cycles. The DLC coating wore through more quickly than TiN. Tetrahedral amorphous carbon, however, showed no visible wear after 5 million cycles.

The abrasion resistance of the tetrahedral amorphous carbon is attributed to its high hardness relative to the DLC and is consistent with a micro-abrasion test performed at Cambridge University [33] .   The results of this test which abraded *TETRABOND*® with 4 micron SiC indicated that it is 100 times more abrasion resistant than TiN.

Conclusions

PVD thin film coatings such as TiN, DLC and tetrahedral amorphous carbon offer a range of properties which show improvement in the corrosion and abrasion resistance of co-based alloys based on electrochemical and pin-on-disk testing.

References

[1] Wehner, G.K., and Anderson, G.S., *Handbook of Thin Film Technology*, McGraw Hill, 1970.

[2] Hakansson, G., "Growth of Compound and Superlattice Thin Films; Effects of Energetic Particle Bombardment," Ph.D. thesis, Department Physics and Measurement Technology, Linkoping University.

[3] Bergman, C., "Ion Flux Characteristics in Arc Vapor Deposition of TiN," Proceedings of the 15th International Conference on Metallurgical Coatings, 1988.

[4] Coll, B.F., and Pellman, M.A., "Metallurgical and Tribological Modifications of Titanium and Titanium Alloys by Plasma Assisted Techniques," Proceedings of the Society for Biomaterials Implant Retrieval Symposium, 1992.

[5] Brown, R., Alias, M.N., and Fontana, R., "Effect of Composition and Thickness on Corrosion Behavior of TiN and ZrN Thin Films," *Surface and Coatings Technology*, 62, 1993, pp. 467-473.

[6] Coll, B.F., et al, "Optimization of Arc Evaporated (Ti,Al)N Film Composition for Cutting Tool Applications," Proceedings of the 18th International Conference on Metallurgical Coatings and Thin Films, 1991.

[7] Chhowalla, M. and Fontana, R., "Deposition of High $Al(Ti_{0.25}Al_{0.75})N$ Films by Cathodic Arc Evaporation," Proceedings of the 22nd International Conference on Metallurgical Coatings and Thin Films, 1995.

[8] Teer, D.G., Jones, A.H.S., and Bellido-Gonzalez, V., "Novel High Wear Resistance DLC Coatings Deposited by Magnetron Sputtering of Carbon," Proceedings of the 24th International Conference on Metallurgical Coatings and Thin Films, 1997.

[9] LoBiondo, N.E., Aharonov, R.R., and Fontana, R., "An Investigation of the Properties of Ti-C:H Films," *Surface and Coatings Technology*, 94-95, 1997, pp. 652-657.

[10] Horsfall, R.; "Tetrahedral Amorphous Carbon Coatings-Properties, Applications & Commercial Successes," Proceedings of the Gorham CVD Diamond & DLC Coatings Conference, March 1996.

[11] Pappas, M.J., Makris, G., and Buechel, F.F., "Comparison of Wear of UHMWPE Cups Articulating with Co-Cr and Ti-N Coated, Titanium Femoral Heads," Trans. 16th SFB, 1990, pp. 36.

[12] Peterson, C.C., Hillberry, B.M., and Heck, D.A., "Component Wear of Total Knee Prostheses Using Ti-6Al-4V, Titanium Nitride Coated Ti-6Al-4V, and Cobalt-Chromium-Molybdenum Femoral Components," *Journal of Biomedical Materials Research*, 22, 1988, pp. 887-903.

[13] Coll, B.F., and Jacquot, P., "Surface Modification of Medical Implants and Surgical Devices Using TiN Layers," *Surface and Coating Technology*, 36, 1988, pp. 867-878.

[14] Coll, B.F., Pellman, M., Souchard, J., and Jacquot, P., *ASTM F-4 Workshop on Ion Implantation of Medical Devices*, May 1991.

[15] Streicher, R.M., Weber, H., Schon, R., and Semlitsch, M., "New Surface Modification for Ti-6Al-7Nb Alloy: Oxygen Diffusion Hardening (ODH)," *Biomaterials*, Vol. 12, 1991.

[16] Davidson, J.A., Mishra, A.K., and Poggie, R.A., "Friction and UHMWPE Wear of Cobalt Alloy, Zirconia, Titanium Nitride, and Amorphous Diamond-Like Carbon Implant Bearing Surfaces," Proceedings of the Fourth World Biomaterials Conference, 1992.

[17] Dahm, K.L., Anderson, I.A., and Dearnley, P.A., "Hard Coatings for Orthopedic Implants," *Surface Engineering*, Vol. 11, No. 2, 1995.

[18] Mauer, A., Brown, S., and Merritt, K., "Effects of Different Surface Treatments on Fretting Corrosion of Ti6Al4V," 38th Proceedings of the Annual Meeting of the ORS, Vol 17, Sec 1, 1992.

[19] Pietra, R., et al, "Titanium Nitride as a Coating for Surgical Instruments Used to Collect Human Tissue for Trace Metal Analysis," *Analyst*, Vol.119, August 1990, pp. 1025-1029.

[20] Wisbey, A., Gregson, P.J., Peter, L.M., and Tuke, M., "Titanium Release from TiN Costed Implant Materials,: *I Mech E*, 1989.

[21] Sella, C. et al, "Corrosion Protection of Metal Implants by Hard Biocompatible Ceramic Coatings Deposited by Radio-Frequency Sputtering," *Clinical Materials*, 5, 1990, pp. 297-307.

[22] Satomi, K., Akagawa, Y., Nikai, H., and Tsuru, H., "Tissue Response to Implanted Ceramic-Coated Titanium Alloys in Rats," *Journal of Oral Rehabilitation*, Vol. 15, 1998, pp. 339-345.

[23] Therin, M., Meunier, A., and Christel, P., "A Histomorphometric Comparison of the Muscular Tissue Reaction to Stainless Steel, Pure Titanium and Titanium Alloy Implant Materials," *Journal of Materials Science in Medicine*, Vol. 2, 1991, pp. 1-8.

[24] Moritmer, E., et al, "Quantitative Evaluation of Bone Apposition and Ingrowth Using Porous Fiber Metal Implants with a Titanium Nitride Surface," Proceedings of the 17th Annual Meeting of the Society for Biomaterials, 1991.

[25] Hayashi, K., et al, "Evaluation of Metal Implants Coated with Several Types of Ceramics as Biomaterials," *Journal of Biomedical Materials Research*, Vol. 23, 1989 pp. 1247-1259.

[26] Van Raay, J.J.A.M., et al, "Biocompatibility of Wear Resistant Coatings in Orthopaedic Surgery in Vitro Testing with Human Fibroblast Cell Cultures," *Journal of Materials Science*, Vol. 6, (2), 1995, pp. 80-84.

[27] Farrari, F., et al, "Metal-ion Release from Titanium and TiN Coated Implants in Rat Bone," *Nuclear Instruments and Methods in Physics Research*, B-79, (1-4), June 1993, pp. 421-423.

[28] Behrndt, H., and Lunk, A., "Biocompatibility of TiN Preclinical and Clinical Investigations," Materials Science and Engineering, A139, 1991, pp. 58-60.

[29] Knotek, O., Loffler, F., and Weitkamp, K., "PVD Coating for Dental Prostheses," Proceedings for the 19th International Conference on Metallurgical Coatings and Thin Films, 1992.

[30] Wisbey, A., Gregson, P.J., and Tuke, M., "Application of PVD TiN Coating to Co-Cr-Mo Based Surgical Implants," *Biomaterials*, Vol. 8, November 1987, pp.477-480.

[31] Takeuchi, M., et al, "The Adhesion of Cathodic Arc Deposited TiN Coatings on Orthopedic Alloys," *Surface Modification Technologies VI*, 1993.

[32] Poggie, R., Correspondence Summarizing Coating Abrasion Tests, 1997.

[33] Horsfall, R., "Tetrahedral Amorphous Carbon Coatings-Properties, Applications & Commercial Successes", Proceedings for Gorham CVD Diamond & DLC Coatings Conference, March 1996.

Alfredo Flores-Valdés, [1] Alfonso H. Castillejos-Escobar, [1] Francisco Acosta-González, [2] José C. Escobedo-Bocardo, [1] and Arturo Toscano-Giles[2]

The Development of Surface Coatings for Co-Cr-Mo Alloys Based on Quaternary AlSiFeMn Intermetallic Compounds

Reference: Flores-Valdés, A., Castillejos-Escobar, A. H., Acosta-González, F., Escobedo-Bocardo, J. C., and Toscano-Giles, A., **"The Development of Surface Coatings for Co-Cr-Mo Alloys Based on Quaternary AlSiFeMn Intermetallic Compounds,"** *Cobalt-Base Alloys for Biomedical Applications, ASTM STP 1365*, J. A. Disegi, R. L. Kennedy, R. Pilliar, Eds., American Society for Testing and Materials, West Conshohocken, PA, 1999.

Abstract: During the development of Co-Cr-Mo as implant alloys, a higher surface hardness is required, in order to avoid the cracks originated by the presence of the chromium or molybdenum carbides in the alloys. Many different methods have been developed, being the surface coating with suitable intermetallic compounds. Surface coatings based on quaternary intermetallic compounds of the AlSiFeMn type have been successfully developed, taking advantage of the fact that chromium diffuses through the boundary layer of the intermetallic alloy, forming a well developed interface. The coatings obtained possess a high microhardness, apart from being continuous and well matched to the base alloy. This paper presents and discusses the methodology that was implemented to produce such coatings at temperatures in the range of 873 to 1123 K, under an applied stress. The coatings were characterized by the use of microprobe analysis, having determined the compositional effects of diffusion of the multiple elements on the coating quality.

Keywords: surface coatings, intermetallics, cobalt-base alloys

Introduction

The synthesis of intermetallic compounds has constituted a field of very active research and a number of manufacturing techniques have now become available, such as segregation from the melt *[1]*, combustion synthesis *[2]*, autopropagation at high temperature *[3]*, spray deposition *[4]*, mechanical alloying *[5]* and others that result from the combination of these techniques.

[1]Researcher, [2]Associated Researcher, CINVESTAV-IPN, Unidad Saltillo, Carretera Saltillo-Monterrey km. 13, 25900 Ramos Arizpe, Coahuila.

Synthesis by autopropagation at high temperature has been used as an alternative to the conventional melting and solidification techniques, since the reaction at the interface propagates very quickly when the necessary energy for the formation of the intermetallic compound becomes available, and defects such as porosity and segregation can be readily eliminated.

The quaternary $Al_8FeMnSi_2$ intermetallic compound has been previously synthesized by a solid state reaction [6], having obtained crystals of uniform morphology and properties. The compound has a density of 3.2 gr/cm^3, an average microhardness of 1047 Vickers, a melting or decomposition temperature of 1069 K. Moreover, it possesses a high resistance to the corrosive attack of sodium hydroxide (NaOH), nitric acid (HNO$_3$), sulfuric acid (H$_2$SO$_4$) and hydrochloric acid (HCl), both diluted and concentrated.

Because chromium is dissolved by the intermetallic compound forming very hard phases, known as superstructures in aluminum alloys [7-9], the idea arose to synthesize it directly over the surface of a Co-Cr-Mo alloy for the development of surface coatings. This idea was supported by the fact that the intermetallic has a great thermodynamic stability at room temperature, and because of its great hardness, it can be used to protect the surface of the alloys against the excessive wear provoked by the chromium and molybdenum carbides that these normally contain.

The solid state reactive synthesis was chosen, because other well known techniques such as packed bed alloying, physical vapor deposition, chemical vapor deposition, vacuum plasma spraying, etc., apart from being expensive, have the limitation that under uncontrolled conditions, they can produce a non uniform matching to the base alloy, great porosity, and microsegregation during the solidification [10-13]. In this sense, the aim of this work is to present the results of an investigation for the possibility of producing surface coatings over a Co-Cr-Mo alloy, based on the $Al_8FeMnSi_2$ intermetallic compound, by a solid state reaction. In the next part of this study the methodology that was implemented to produce the $Al_8FeMnSi_2$ intermetallic surface coatings over the ASTM F75 wrought alloy will be presented.

Experimental Procedure

The process to produce the $Al_8FeMnSi_2$ intermetallic surface coatings over a Co-Cr-Mo alloy was based on the following reaction [14]:

$$8Al + 2Si + Fe + Mn = Al_8FeMnSi_2 = f(T) \tag{1}$$

where:
T is the temperature at which the reaction is carried out.

The chemical free energy of formation of the $Al_8FeMnSi_2$ compound is then given by:

$$\Delta G_f^0 = -RT \ln a_{Al_8FeMnSi_2} \tag{2}$$

where:

ΔG_f = free energy of formation of the intermetallic compound (J/mol)

R = universal constant of gases (1.98 J/mol-K)

T = temperature (K)

$a_{Al_8FeMnSi_2}$ = activity of the intermetallic compound

if the reactants are thermodynamically unstable and the formation of the intermetallic compound proceeds to a certain degree of efficiency. From a previous work [14] it was determined that the expression for the free energy of formation of the intermetallic compound is:

$$\Delta G_f^0 = -186\,238 + 59.63T \text{ (J/mol)} \tag{3}$$

which indicates, by its negative value, the thermodynamic stability of the compound, and therefore the reliability of its formation. The efficiency of the reaction depends, however, on the particle morphology and size distributions of the reactants and also on the physical and chemical properties of the contact surfaces. The chemical process indicated by Eq. 1 may be carried out, therefore, if the ambient temperature, pressure of compaction, class of atmosphere, and initial concentration of the solid reactants are selected appropriately. The stoichiometry of the intermetallic compound was previously determined by chemical analysis of a number of particles of the compound synthesized by segregation from the melt [14]. Then, the measured chemical composition of the intermetallic compound was 57.98 wt % Al, 13.37 wt % Fe, 15.40 wt % Mn, and 13.25 wt % Si, corresponding to the formula $Al_8FeMnSi_2$. It must be stated that chromium can be dissolved by the intermetallic compound, so the chemical composition of the intermetallic could be modified, and its corresponding formula would be of the type $Al_8FeMnCrSi_2$, as it appears in different scientific papers [15,16].

   In the present work, the solid state reactive synthesis of the intermetallic compound was performed using commercially pure Al, Mn, Fe and Si, and as a substrate, a Co-Cr-Mo alloy whose chemical composition corresponded to the ASTM F75 wrought alloy. The purity of the reactants was determined by atomic absorption spectrophotometry and inductively coupled argon plasma spectrometry. The purity reported was 99.84 wt % for aluminum, 99.56 wt % for iron, 99.79 wt % for manganese and 98.82 wt % for silicon. In the case of silicon, manganese and aluminum, iron was the main impurity, so for the calculation of the stoichiometric amount required of this element, its content in these other elements was taken into account.

   The process to develop the surface coatings over the samples of the Co-Cr-Mo alloy was performed as follows: discs of 3.0 mm in thickness of this alloy were cut off from a bar 25.4 mm in diameter, activating their surface by the action of sand blasting, using silicon dioxide (SiO_2) 0.3 mm in diameter as sandblasting medium. Then, 1.5 grams of ultrasonically mixed powders consisting of Al, Si, Fe, and Mn in the stoichiometric relationship given by the formula of the $Al_8FeMnSi_2$ intermetallic compound were pressed over the surface of each sample, at a predetermined stress of 0.5 kg/mm². This stress was enough to produce a layer of 1 mm high. The size of the elemental powders was fixed at the value of 0.250 mm. Then, each disc was positioned inside an isolated chamber attached to a machine that was able to perform a constant compressive action at high temperature, and under a protective atmosphere. The compression was transmitted through two bars made of a molybdenum TZM alloy, whose nominal composition is 0.5Ti, 0.18 Zr, 0.015C, Mo balance. As a protective atmosphere ultra high purity argon was used. During the tests, the flow of argon was maintained at the fixed value of 0.22 lt/min. The parameters that were selected for this study were temperature and stress; the values of these parameters are as follows: the temperatures chosen were 873, 923, 973, and 1023 K. Meanwhile, the stresses selected were 1.0, 1.25, and 1.5 kg/mm^2. In all cases, the time of reaction was constant, at the fixed value of 120 minutes. The experiments were statistically designed to fit a random factorial block model, with three repetitions per experiment.

Results and Discussion

Characterization of the Reaction Products

After performing each test, the discs obtained were removed from the compression machine and prepared for quantitative metallographic measurements. These measurements were carried out using both optical and scanning electron microscopy. The solid cylindrical samples obtained after each one of the experiments were sectioned longitudinally through their centerline and mounted in bakelite. The samples were then prepared metallographically until reaching a mirror-like finish, and then examined under an optical microscope equipped with an image analysis system. This arrangement was used to determine the quality and thickness of the coatings produced. As an example, Figure 1 and Figure 2 show different aspects of the coatings obtained, where the good matching of the intermetallic over the surface of the Co-Cr-Mo alloy is evident.

Next, semi-quantitative X-ray energy dispersive spectroscopy (EDS) and scanning electron microscopy (SEM) were used to determine the chemical composition and microstructure of the coatings produced. Taking as an example the photomicrographs of Figures 1 and 2, and according to the chemical microanalysis reported, three different zones can be distinguished. The outer zone is composed mainly of the $Al_8FeMnSi_2$ intermetallic compound. The spectrum of Figure 3 corresponds to the composition of the intermetallic in the coating, which is comparable to the spectrum of the pure intermetallic, as shown in Figure 4. Meanwhile, the inner surface corresponded to the cobalt base alloy. There is an interface between the cobalt base alloy and the intermetallic compound, as can be distinguished in the micrographs, characterized by the presence of all elements participating in the surface reaction. The corresponding EDS spectrum is shown in Figure 5.

On the other hand, microhardness values were obtained for each sample, measuring from the outer surface through the inner surface of the samples. For these determinations, a load of 200 grams-force was used. Table 1 shows the results obtained for the samples indicated, where it is evident there are differences between the microhardness of the different layers.

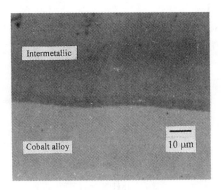

Figure 1- *Surface Coating Produced at 923 K and 1.25 kg/mm²*

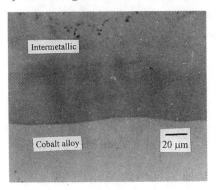

Figure 2- *Surface Coating Produced at 873 K, and 1.0 kg/mm²*

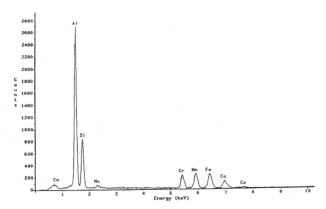

Figure 3- *EDS Spectrum of the Al₈FeMnSi₂ in the Coating*

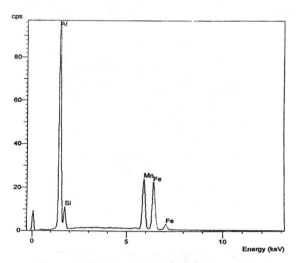

Figure 4- *EDS Spectrum of a Reference Sample*

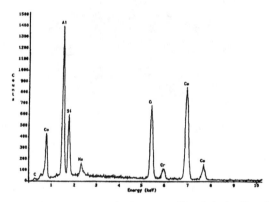

Figure 5- *EDS Spectrum at the Intermetallic-Cobalt Alloy Interface*

From the results of this work, it is evident that good surface coatings can be obtained, as a consequence of the fair matching developed at the intermetallic-cobalt base alloy interface. Table 2 shows the results obtained for all the conditions investigated. From the values reported in this table, it can be seen that for all the conditions studied the development of the intermetallic coating-cobalt base alloy is feasible. Nevertheless, the higher the temperature and the stress imposed, the better the continuity of the intermetallic produced.

Table 1.- *Microhardness Measurements at the Different Zones of Reaction of the Samples Indicated.*

Sample Obtained at:	Intermetallic Microhardness (HV)	Cobalt Alloy Microhardness (HV)	Interface Microhardness (HV)
873 K, 1.00 kg/mm^2	873	312	389
873 K, 1.25 kg/mm^2	913	316	367
873 K, 1.50 kg/mm^2	927	303	392
923 K, 1.00 kg/mm^2	936	306	401
923 K, 1.25 kg/mm^2	868	30	395
923 K, 1.50 kg/mm^2	892	332	368
973 K, 1.00 kg/mm^2	1073	296	399
973 K, 1.25 kg/mm^2	1036	288	401
973 K, 1.50 kg/mm^2	1051	306	416
1023 K, 1.00 kg/mm^2	1100	283	376
1023 K, 1.25 kg/mm^2	976	296	394
1023 K, 1.50 kg/mm^2	1083	313	399

This is a consequence that the process is of the diffusional type, and is affected greatly by the temperature and the stress applied. The fact that there is a chemical reaction during the synthesis of the intermetallic compound over the surface of the cobalt base alloy makes it difficult to explain the kinetics of the formation of the intermetallic compound and its good matching to the base alloy.

Previous studies on the formation of the $Al_8FeMnSi_2$ quaternary intermetallic compound by solid state reaction synthesis have shown that the interface of reaction among the elements propagates quickly once a certain level of energy is reached [6, 17]. During this reaction synthesis process, the formation of the intermetallic occurs around 998 K, with the application of a constant external pressure. In the case of the obtaining a coating over the surface of the cobalt base wrought alloy ASTM F75, the fact that chromium can diffuse through the intermetallic compound, forming part of it, was considered. The process was complex, because on one hand, the formation of the intermetallic compound must occur. On the other hand, obtaining a coating of this compound over the surface of the cobalt-base alloy must also occur. The constant external stress applied during the experiments was responsible in great part for the great continuity of the layer attained, because the contact area of the reaction was improved.

The nature of the coatings obtained was investigated using the scanning electron microscope to perform "line scan" patterns. Table 3 shows the penetration distances of each element participating in the reaction, that have been determined for the samples obtained at 1023 K, under an applied stress of 1.5 kg/mm^2. The line of scanning was chosen as 0.137 mm, starting from the outer surface of the sample, as indicated in the photomicrograph of Figure 6.

Table 2.- *Optical Microscopy Measurements of the Samples Indicated.*

Sample Obtained at:	Thickness of the Layer of Intermetallic (μm)	Thickness of the Cobalt Alloy-Intermetallic interface (μm)
873 K, 1.00 kg/mm^2	75.62	4.75
873 K, 1.25 kg/mm^2	77.23	3.97
873 K, 1.50 kg/mm^2	81.25	4.35
923 K, 1.00 kg/mm^2	86.44	5.23
923 K, 1.25 kg/mm^2	86.76	5.66
923 K, 1.50 kg/mm^2	89.39	5.44
973 K, 1.00 kg/mm^2	100.00	12.56
973 K, 1.25 kg/mm^2	102.44	13.00
973 K, 1.50 kg/mm^2	106.33	16.18
1023 K, 1.00 kg/mm^2	110.00	20.25
1023 K, 1.25 kg/mm^2	108.67	21.34
1023 K, 1.50 kg/mm^2	111.34	25.45

This total distance was chosen to include the intermetallic compound layer, the layer of reaction at the intermetallic-cobalt base alloy interface, and a layer of 0.06 mm of the base alloy itself. The percentage of X-rays produced by the elements in the different layers gave an idea of the composition in these zones, so these values can be used in a semi-quantitative way to assess the chemical composition along the line of scanning.

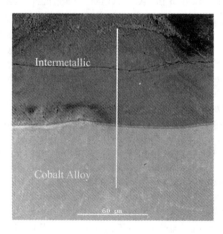

Figure 6- Size of the Line of Scanning Practiced in a Sample Produced at 1023 K, and 1.50 kg/mm^2

Table 3.- *Penetration Distances Measured for the Elements Participating in the Formation of a Coating at 1023 K, and 1.5 kg/mm^2 of Stress.*

Element	Distance from Outer Surface (mm)	% of Maximum Intensity of the X-Rays Produced by the Element
Aluminum	0 - 0.038	100
	0.038 - 0.073	68
	0.073 - 0.076	60
	0.076 - 0.084	52
	0.084 - 0.137	0
Silicon	0 - 0.071	100
	0.071 - 0.077	70
	0.077 - 0.082	43
	0.082 - 0.137	0
Manganese	0 - 0.017	100
	0.017 - 0.037	60
	0.037 - 0.047	36
	0.047 - 0.137	0
Iron	0 - 0.034	100
	0.034 - 0.036	66.67
	0.036 - 0.041	50
	0.041 - 0.045	33
	0.045 - 0.050	16
	0.050 - 0.137	0
Chromium	0 - 0.038	0
	0.038 - 0.041	20
	0.041 - 0.072	40
	0.072 - 0.077	52
	0.077 - 0.137	100
Cobalt	0 - 0.035	0
	0.035 - 0.050	20
	0.050 - 0.070	40
	0.070 - 0.073	54
	0.073 - 0.077	72
	0.077 - 0.137	100
Molybdenum	0 - 0.038	0
	0.038 - 0.067	45
	0.067 - 0.076	60
	0.076 - 0.079	90
	0.079 - 0.137	100

As can be seen from the data reported in Table 3, the length of the reaction zone is approximately 0.040 mm, that is the region where interdiffusion among elements occurred. Out of this limit, either the pure intermetallic compound or the cobalt base alloy exist. From these results it can be concluded that aluminum, silicon, cobalt, chromium, manganese, and molybdenum helped to form a small region of continuity. This finding is slightly different than the original proposition, because it was believed that only chromium would take an important role in the interface formation. Because of the great microhardness achieved by the coatings produced, it is suggested to conduct experimental work, in order to investigate the biocompatibility and wearing behavior of the coatings obtained.

Conclusions

1.-   The development of surface coatings over the ASTM F75 cobalt base alloy based on the $Al_8FeMnSi_2$ intermetallic compound has been investigated, and it was found that this process is feasible.

2.-   The best coatings were obtained when the elemental powders of the intermetallic compound reacted at a temperature of 1023 K, under a constant stress of 1.5 kg/mm^2, during a period of 120 minutes.

3.-   The main characteristics of the coatings produced were a thickness on the order of 0.101 mm as average, a great uniformity through the contact area, an absence of porosity and segregation, and a great microhardness, in the order of 1100 Vickers.

4.-   It is believed that these characteristics were achieved by the fact that chromium, cobalt, and molybdenum diffuse, as together with Al, Si, and Mn, through the boundary layer, developing a zone of reaction of nearly perfect matching

References

[1] Brown, A. and Westbrook, J. H., "Formation Techniques for Intermetallic Compounds, " *Intermetallic Compounds*, Edited by J.H. Westbrook, John Wiley and Sons, Inc., N.Y., 1967, pp. 315-336.

[2] Bockowski, M. et al., "Combustion Synthesis of Aluminum Nitride Under High Pressure of Nitrogen and Nitrogen-Argon Mixtures, " *Journal of Materials Synthesis and Processing*, Vol. 5, No. 6, 1997, pp. 449-458.

[3] Capaldi, M. J. and Uwakweh, O.N.C., "Production and Characterization of TiC Containing Materials by Self-Propagating High Temperature Synthesis, " *Journal of Materials Synthesis and Processing*, Vol. 4, No. 4, 1996, pp. 245-253.

[4] Stelzer, N.H.J. and Schoonman, J., "Synthesis of Terbia-Doped Yttria-Stabilized Zirconia Thin Films by Electrostatic Spray Deposition, " *Journal of Materials Synthesis and Processing*, Vol. 4, No. 6, 1996, pp. 429-438.

[5] Jordan, A. and Uwakweh, O.N. C, "The Study of Mechanically Alloyed Fe-Zn-Si Intermetallic Phases, " *Journal of Materials Synthesis and Processing*, Vol. 5, No. 2, 1997, pp. 169-180.

[6] Castillo-Martínez, D., Bachelor's Thesis, "Síntesis de Compuestos Intermetálicos Tipo AlFeMnSi, " Autonomous University of Coahuila, Saltillo, Coahuila, México, December 1996.

[7] Yaneva, S., Stoichev, N., Kamenova, Z., and Budurov, S., "Quaternary Iron-Containing Phases in Al-Si Cast Alloys, " *Zeizchrift fur Metallkunde*, Vol. 75, No. 5, May 1984, pp. 395-398.

[8] Munson, D., "A Clarification of the Phases Occurring in Aluminium-Rich Aluminium-Iron-Silicon Alloys, with Particular Reference to the Ternary Phase alpha-AlFeSi, " *Journal of the Institute of Metals*, Vol. 95, 1967, pp. 217-219.

[9] Tamminen, J. and Backerud, L., "Solidification Characteristics of Some Al-Base Alloys, " (PROC) Solidification Processing, Sheffield, England, September 21-24, 1987, pp. 435-438.

[10] Gale, W.F. and King, J.E., "Microstructural Development in Aluminide Diffusion Coatings on Nickel-Base Superalloy Single Crystals, " *Surface and Coatings Technology*, Vol. 54/55, Nos. 1-3, November 1992, pp. 8-12.

[11] Salvador-Fernández, J.C. and Ferreira, M.G.S. , "Corrosion Behavior of Physically Vapor Deposited Al-Zn Coatings on 7075 Aluminum Alloy, " *Surface and Coatings Technology*, Vol. 53, No. 1, July 1992, pp. 99-100.

[12] Bartsch, K., Leonhardt, A. and Wolf, E., "Deposition of Multilayer Hard Coatings Using Kinetically Controlled Chemical Vapor Deposition Processes, " *Surface and CoatingsTechnology*, Vol. 54/55, No. 1, November 16, 1992, pp. 193-197.

[13] Moreau, C., Lamontagne, M., and Cielo, P., "Influence of the Coating Thickness on the Cooling Rates of Plasma-Sprayed Particles Impinging on a Substrate, " *Surface and Coatings Technology*, Vol. 53, No. 2, September 1992, pp. 107-114.

[14] Flores-Valdés, A., "Aluminum Scrap Purification Through the Segregation of Intermetallic Phases, " Ph.D. Thesis, University of Mining and Metallurgy, Krakow, Poland, March 1994.

[15] Flores-Valdés, A. et al., "Differential Thermal Analysis of $Al_8FeMnSi_2$ Intermetallic Phase Particles, " *Scripta Metallurgica et Materialia*, Vol. 30, 1994, pp. 435-439.

[16] Flores-Valdés, A. et al., "Characterization of $Al_8FeMnSi_2$ Phase Crystals Grown From Al-Si-Fe-Mn Melts, " *Journal of Materials Synthesis and Processing*, Vol. 2, No. 1, 1994, pp. 51-55.

[17] Flores-Valdés, A. et al., "A Kinetic Study on the Nucleation and Growth of the $Al_8FeMnSi_2$ Intermetallic Compound for Aluminum Scrap Purification, " *Intermetallics*, Vol. 6, 1998, pp. 217-227.

Clinical Experience

Pat Campbell,[1] Harry McKellop,[2] Rina Alim,[3] Joseph Mirra,[4] Steven Nutt,[3] L. Dorr[5] and H. C. Amstutz[6]

Metal-On-Metal Hip Replacements: Wear Performance and Cellular Response To Wear Particles

---

Reference: Campbell, P., McKellop, H., Alim, R., Mirra J., Nutt, S., Dorr, L., and Amstutz, H. C., **Metal-On-Metal Hip Replacements: Wear Performance and Cellular Response To Wear Particles,** *Cobalt-Base Alloys for Biomedical Applications, ASTM STP 1365,* J. A. Disegi, R. L. Kennedy, and R. Pilliar, Eds., American Society for Testing and Materials, West Conshohocken, PA, 1999.

Abstract: This study examined the clinical wear performance of second-generation metal-on-metal (M-M) total hips and evaluated the cellular reaction to, and characteristics of, the wear particles. Twenty cobalt chrome (CoCr) M-M THRs were studied, including ten Metasul conventional total hips (THRs), and ten McMinn surface replacements (SRs). As in the first-generation M-M hips, wear ranged from undetectable to 32 µm in the SR femoral shells (ave 15.7 ± 14 µm) and from undetectable to 19 µm in the THR balls (ave 10.1 ± 7 µm). Self-polishing of third body scratches was evident. Two THRs exhibited clusters of micropits, which had been described on first-generation M-M hips but these did not appear to correlate with accelerated wear. CoCr particles were either dense or amorphous and were mostly in the nanometer size range. There were generally fewer macrophages in the tissues than seen with M-PE THRs, and there was a more inflammatory response to bone cement particles compared with metal particles. Nevertheless, the long-term response to these very small CoCr particles should be monitored.

Keywords: cobalt chrome, wear particles, wear, total hip arthroplasty, histology

---

[1] Director, Implant Retrieval Laboratory, Joint Replacement Institute, Orthopaedic Hospital, 2400 S. Flower St., Los Angeles, CA 90007.
[2] Director of Research, Orthopaedic Hospital, 2400 S. Flower St., Los Angeles, CA 90007.
[3] MS candidate and Professor, University of Southern California, Department of Materials Science, Los Angeles, CA 90007.
[4] Pathologist in Chief, Orthopaedic Hospital, 2400 S. Flower St., Los Angeles, CA 90007.
[5] Director and Professor of Orthopaedic Surgery, University of Southern California, Department of Orthopaedic Surgery, Los Angeles, CA 90033.
[6] Medical Director, Joint Replacement Institute, Orthopaedic Hospital, 2400 S. Flower St., Los Angeles, CA 90007.

Introduction

Approximately 250 000 total hip replacements are performed in the United States every year, with an increasing proportion of these being done in younger patients. Unfortunately, the annual rate of revision total hip surgery is increasing and together, total hip and knee revisions accounted for about 10% of all hip and knee replacement cases in the period 1993 - 1994[1]. The reasons for failure have been the focus of intensive research and there is now a consensus of opinion that ultra high molecular weight polyethylene (UHMWPE) wear debris is the primary cause. In 1995, a symposium sponsored by the American Academy of Orthopaedic Surgeons noted that wear is an important factor that affects the fixation and durability of the implants, and that wear related problems may be responsible for about 9% of all joint replacement surgeries each year *[1]*.

After extensive research into the characteristics of UHMWPE wear particles, it is now becoming clear why they cause bone loss and aseptic loosening. UHMWPE components of total hip replacements wear on the order of 50 -100 mm^3 per year [17]. Since the majority of the particles are submicron in size, this volume of wear comprises billions of submicron wear particles that are in the size range for phagocytosis by macrophages *[2-4]*. The material is indigestible within the cells, and the elongated shape of many of the particles may be an irritant within the cell. For these reasons, following ingestion of the particles, the macrophages become activated and release numerous inflammatory cytokines that can stimulate osteoclast activity, as well as products that may directly dissolve bone *[5-10]*. The resulting loosening is one of the major causes of total hip replacement failure. Considering that the economic cost, the length of hospital stay and the incidence of complications are significantly increased with revision surgery, efforts to reduce wear-related revisions are receiving widespread attention within the orthopaedic community.

Alternative Bearings for Total Hip Replacement

Metal-on-metal (M-M) THRs were in clinical use in the early 1960's but were mostly replaced by metal-UHMWPE designs. Recently, there has been a resurgence in interest in M-M bearings to avoid the problems associated with UHMWPE wear debris. Retrieval analyses of the early components have been carried out and hip simulator tests of second generation components are being conducted. The results from these studies show that M-M THRs wear approximately <1 - 6mm^3/yr per year *[11,12, 17]*. The tissues around failed M-M joints are characterized by fibrous tissues rather than the histiocytic granulomas that are typical around failed metal-UHMWPE joints and osteolysis has not been a major cause of failure in M-M joints *[13,14]*.

The Biological Response to Wear Particles

All joint replacements will produce wear particles as an inevitable consequence of their use in vivo. The tissue response to the particles will be determined by the composition, size, shape and number of the particles *[1, 27]*. While the characteristics of UHMWPE particles and the tissue response to them have been extensively studied, the same is not true for metal particles. The characteristics of the metal particles will be determined by the wear processes acting on the components, but metal particles may also undergo changes once they are exposed to body fluids or intracellular enzymes. There are concerns that metal particles from cobalt chrome (CoCr) alloy implants may cause allergic responses, toxic responses or even cancerous changes in the tissues [15,16].

The aims of this study were to 1) examine the clinical wear performance of second generation M-M THRs 2) study the histological appearance of the periprosthetic tissues as an indication of the biological response to the particles and 3) characterize the wear particles produced in M-M THRs.

Materials and Methods

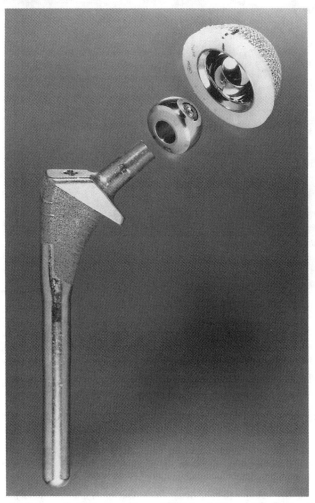

Figure 1 - *Intermedics Titanium Stem with Metasul Wrought CoCr Bearings*

Periprosthetic tissues from ten hips with Metasul™ THRs (Sulzer Orthopaedics, Switzerland) (Figure 1) and ten with McMinn surface replacements (SRs) (Corin, UK) (Figure 2), were studied. Reasons for failure are given in Table 1. The THR bearings were made from wrought CoCrMo alloy while the SRs were made from cast CoCrMo alloy.

Figure 2 - *McMinn Surface Replacement with Cast CoCr Bearings*

Seven of the femoral stems of the THRs were made of titanium alloy while the remainder were made of CoCr alloy. The majority of the SR components were fixed with polymethyl methacrylate (PMMA) bone cement while seven of the THR components were cemented on at least one side. One SR was cut to allow the interfaces between the femoral head bone and the implant to be examined.

Assessment of in Vivo Wear

The wear of eight THRs and eight SRs retrieved after nine months to 6.5 years of use was characterized using laser profilometry, (Perthometer, Mahr/Feinpruf, Gottingen, Germany) a coordinate measuring machine (CMM, BMT 504, Mitotoyo, Aurora, IL) and scanning electron microscopy (SEM, Zeiss, West Germany), as previously described [17].

Table 1 - *Clinical Details of Cases Included in this Study*

Case	Age	Time in vivo, months	Gender	Weight, kg	Activity	Cement Fixation	Why Revised	Ball Size, mm
SR1	66	19	M	105	High	A-yes F-yes	Autopsy	52
SR2	57	26	F	86	Low	A-yes F-yes	Acet Loose	44
SR3	56	15	F	86	Low	A-yes F-yes	Acet Loose	44
SR4	63	9	M	93	Low	A-yes F-yes	Fracture	52
SR5	42	26	M	84	Low	A-yes F-yes	Debonding	48
SR6	50	7	M	99	Low	A-no F-yes	Acet Loose	52
SR7	28	25	F	114	Low	A-yes F-yes	Acet Loose	44
SR8	32	12	F	81	Low	A-no F-yes	Dislocation	44
SR9	26	32	F	N/k	Mod	A-yes F-yes	Acet Loose	44
SR10	56	25	M	100	Low	A-yes F-yes	Infection	44
THR1	57	36	M	100	Low	A-no F-yes	Broken Stem	28
THR2*	52	34	F	64	High	A-yes F-no	Dislocation	28
THR3*	72	26	M	N/k	Mod to high	A - no F - no	Autopsy	28
THR4	75	48	F	90	Low	A -no F - no	Acet Loose	28
THR5	84	27	M	83	Low to mod	A-yes F-yes	Autopsy	32
THR6*	67	28	M	N/k	N/k	A -no F - no	Fem Osteolysis	32
THR7*	69	58	F	N/k	N/k	A-yes F-no	Dislocation	28
THR8*	84	19	M	69	Low to mod	A-yes F-no	Autopsy	28
THR9*	54	31	M	113	Mod	A-no F-no	Fem Loose	28
THR10*	52	14	M	N/k	N/k	A-yes F - no	Infection	28

* Stems made of Ti alloy.

N/k = not known.

A = acetabular , F = femoral.

Assessment of Tissue Response to Wear Particles

Representative samples of tissue from the capsule and interfacial membranes were processed into paraffin wax and sections were cut at 6 μm and stained with haematoxylin and eosin for light microscopic analysis. A semi-quantitative rating scheme was used to assess the number of macrophages, giant cells, and visible metal particles, as well as the appearance of the macrophages (slate blue, dusty or black), and the extent of necrosis [13]. Ten fields were examined from each tissue section and the results were presented as the mode of the observations.

Tissue samples approximately 100-150 mg in size were taken from the same periprosthetic sites used for tissue sections from four SRs and five THRs. These specimens were chosen based on the histological appearance of metal and/or the grey discoloration of the tissue, since a relatively high proportion of metal is required for successful isolation. The metal particles were isolated from the tissues using the technique of Doorn et al. [18]. A 100 mL aliquot of the isolated particle pellet was sprayed onto carbon coated, formvar, copper mesh grids (Ted Pella, Redding, CA) using a glass nebulizer (Ted Pella, Redding, CA) in preparation for TEM analysis. The samples were examined in a transmission electron microscope (TEM 420, Philips, The Netherlands), using both bright and dark field techniques at 120 kV. The bright field and dark field images were obtained by selecting either the direct beam or the diffracted beams in the SAD pattern at maximum tilt with the objective aperture, respectively.

Assessment of CoCr Particle Morphology

Elemental analysis was performed in the TEM using an X-ray detector (Kevex Sigma, San Carlos, CA). For the purposes of this study, a wear particle was defined as a polycrystalline feature within the TEM field that appeared to have an enclosing boundary. EDXA verification was performed before the particle was photographed for further analysis. Photographs of the metal particles were transferred to a computer based image analysis system (Image One West, Chester, PA) and the particle sizes were measured.

Results

In Vivo Wear

The results of the laser profilometry and the CMM measurements are shown in Table 2. The diametral clearances averaged 115 μm in the THRs and 288 μm in the SRs. The wear depth on the eight THRs ranged from not detectable (<1 μm) to 9 μm. The wear of three femoral balls and four cups of the SRs could not be reliably distinguished from the initial out-of-roundness. In the remaining SRs, the wear depth ranged from not detectable to 20 μm. The profilometry values were reflected in the SEM observations of the surfaces.

The non-worn surfaces in all cases exhibited prominent carbide bumps and light residual polishing scratches. In the worn zones, the residual polishing scratches had been smoothed out, and there were larger scratches from third-body abrasion which also had

Table 2 - *Component Wear Results: CMM and Profilometry*

Case	Nominal Diameter (mm)	Dome Ra (μm)	Dome Rmax (μm)	Max. Deviation (μm)	Min. Deviation (μm)	Diametral Clearance	Wear Depth (μm)
SR1 ball	52	0.09	2.58	9	12	260	32
SR1 cup	52	N/A	N/A	15	19		N/D
SR3 ball	44	0.38	9.13	5	5	298	9
SR3 cup	44	N/A	N/A	21	28		N/D
SR4 ball	52	0.12	2.67	2	6	-74	6
SR4 cup	52	N/A	N/A	5	6		6
SR5 cup	48	N/A	N/A	8	12	N/A	N/M
SR6 ball	52	0.13	2.37	89	24	390	N/D
SR6 cup	52	N/A	N/A	7	5		N/D
SR7 cup	44	N/A	N/A	6	7	N/A	N/M
SR10 ball	44	0.17	3.10	4	4	228	N/D
SR10 cup	44	N/A	N/A	9	5		N/D
THR1 ball	28	0.11	1.66	4	3	N/A	N/M
THR2 ball	28	0.09	0.75	2	2	120	15
THR2 cup	28	N/A	N/A	2	4		N/M
THR3 ball	28	0.13	3.41	3	3	103	19
THR3 cup	28	N/A	N/A	3	3		9
THR4 ball	28	0.20	2.28	2	2	120	17
THR4 cup	28	N/A	N/A	3	3		12
THR5 ball	32	0.17	3.35	3	2	152	N/M
THR5 cup	32	N/A	N/A	3	2		N/M
THR 6 ball	28	0.11	1.69	3	3	100	N/M
THR 6 cup	28	N/A	N/A	4	4		N/M
THR7 ball	28	0.23	4.41	2	2	100	7
THR7 cup	28	N/A	N/A	3	3		N/M
THR8 ball	28	0.17	1.88	2	4	120	3
THR8 cup	28	N/A	N/A	2	2		N/M
THR9 ball	28	0.15	2.56	2	2	96	4
THR9 cup	28	N/A	N/A	2	3		4
THR10 ball	28	0.12	1.93	3	3	112	6
THR10 cup	28	N/A	N/A	3	4		N/M

N/A = not available, e.g., clearance requires both components, and perthometry cannot be performed on cups.

N/D = not determined, due to high out of roundness.

N/M = not measurable, due to low wear such that wear could not be distinguished from the original form of the implant.

been partially polished out over time. In the THRs, most of the carbides had been worn down to the level of the surrounding matrix, and some were <u>below</u> the matrix, forming flat-bottomed carbide depressions less than 1 μm deep. In addition, two of the THRs (THR #4, THR #10) exhibited clusters of micropits in the main bearing area. These measured 1 to 3 μm in diameter and <1 μm deep. These features did not appear to be associated with high wear as THR #4 had a wear depth of 17 μm on the ball and 12 μm on the cup while THR #10 had a wear depth of 6 μm on the ball and the wear depth was too small to be determined on the cup.

Tissue Response to Wear Particles

The tissues were grossly grey-black (metallosis) in two of the SRs and three of the THRs. In all but one SR case (SR1) this was due to titanium particles released due to impingement of the femoral stem against the socket, debonding of the porous coating or stem breakage. In the remaining case, the discoloration was due to CoCr particles from the wear-in phase of the components. There was light grey discoloration in one of the SRs and three of the THRs due to component loosening (i.e. from the abrasion of the loose stem or cup within the bone or debonded cement mantle) or to wear from the third body particles generated by the loosened implant. The histological appearance reflected the gross appearance. In the normal appearing tissues, only small numbers of macrophages with visible metal particles were noted, usually along the outer edges of the tissue or close to subsurface vessels. In the metallosis cases, macrophages filled with black, metallic particles were concentrated around the edges, and sometimes extended deeper into the tissue. The light grey tissues contained macrophages with a dusty or slate blue appearance.

There was considerable variation within each tissue sample for the numbers of macrophages, giant cells and wear particles (Table 3). Generally, there were fewer macrophages and wear particles in the tissues than is typically seen in tissues around metal-polyethylene joints. Metal particles were seen in small numbers in each case, but PMMA particles, as evidenced by the characteristic "holes" in cells or tissues, were rarely seen in the cemented cases. Polyethylene particles, presumably from the socket inner lining, were seen in small numbers in most of the Metasul cases. The metal particles ranged in size from barely visible to several microns in diameter and appeared as dark grains or needles. The appearance of slate-blue histiocytes suggested the presence of fine, intracellular metal. Polyethylene particles were mostly needle shaped, 2 to 5 μm in size and located in macrophages, although some larger particles were occasionally seen within giant cells in the cases that had undergone dislocation and rim impingement.

Small areas of histiocytic granulomas were noted in five of the SRs in association with metal and/or PMMA wear debris. The femoral head in an SR that had been sectioned was intact with minimal interfacial membrane formation and no osteolysis of the cancellous bone. There were no granulomas in the Metasul group tissues, including the one case revised for distal femoral osteolysis. The histology of that lesion was inconsistent with wear-induced osteolysis, and was not consistent with infection. Small, focal accumulations of giant cells and macrophages containing polyethylene particles were seen in half of the Metasul cases. Extensive necrosis was noted in one of the SRs and one THR, with smaller focal necrotic lesions in four of the remaining SRs and three of the remaining THRs. There was no clear association between necrosis and metal wear particles or the gross appearance of the tissues.

Table 3 - *Results of the Semi-Quantitative Rating of the Histological Features in the Periprosthetic Tissues*

Case	Tissue	Macro-phages	Giant Cells	Metal Particles	Metal Histiocyte	PMMA	Necrosis
SR1	Ant Capsule	2, 0	0	0	0	0	0
	Capsule-2	1	0	1	1	0	0
SR2	Ant Capsule	0	0	0	0	0	0
	Post Inf Capsule	0	0	0	0	0	0
SR3	Ant Inf Capsule	N/A	N/A	N/A	N/A	N/A	3
SR4	Acet Membrane	0	0	0	0	0	0
	Sup Capsule	1	0	1	0	0	0
SR5	Post Sup Capsule	1	0	2	2	0	0
	Acet Membrane	2	1	1	0	1	0
SR6	Inf Capsule	3	0	0	0	0	0
	Acet Membrane	0	0	1	N/A	0	0
SR7	Ant Inf Capsule	2	0	1	2	0	0
	Acet Membrane	2	0	2	2	0	0
SR8	Ant Sup Cyst	1	0	2	2	0	0
	Post Sup Capsule	0	0	0	0	0	0
SR9	Ant Capsule	3	0	2	2	0	0
	Unlabeled	3	0, 1	2	2	0	0
SR10	Unlabeled-1	0	0	0	0	0	0
	Unlabeled-2	N/A	N/A	N/A	N/A	N/A	3
THR1	Unlabeled-1	0	0	0	0	0	0
	Unlabeled-2	1	0	1	0	0	0, 2
THR2	Capsule	2	0	1	2	0	0, 1
	Acet Membrane	2, 3	0	3	3	0	0
THR3	Capsule	1	0	2	2	0	0
THR4	Capsule	N/A	N/A	N/A	N/A	N/A	3
THR5	Med Capsule	0	0	0	0	0	0
	Inner Interface	0	0	0	0	0	0
THR6	Capsule	0	0	0	0	0	3
	Distal Femur	1	0	0	0	0	0
THR7	Unlabeled	0, 1	0	0	0	0	0
THR8	Capsule-1	1	0	0	0	0	0
	Capsule-2	0	0	0	0	0	0
THR9	Ant Membrane	2	0	0	1	0	1
	Sup Capsule	0	0	0	0	0	2
THR 10	Capsule	0	0	0	2	0	0

N/A = Cell details not determinable due to necrotic changes.

Morphology of CoCrMo Particles

Metal particles were readily apparent, either singly or in clusters on the TEM grids. More particles were found in the tissues that were grossly grey. Two consistent particle forms were seen in both groups. One was a dense, elongated form which usually had a defined edge with crystalline areas dispersed throughout the particles. These were generally 30 to 150 nm in size and had an EDXA profile of high cobalt and chromium, (Fig. 3a and b). The second, and most abundant, form had less defined edges with a non-homogeneous texture and density, and was described as amorphous. These particles were generally up to several hundred nanometers in size, and seemed to be composed of smaller particles that had combined to form larger particles. Their EDXA profile indicated high chromium and oxygen but low cobalt (Fig. 4a and b). The particle size was comparable for the SRs and THRs (Table 4), with 92% and 72% of the particles from SRs and THRs being less than 200 nm respectively.

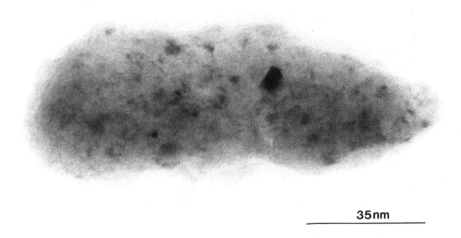

35nm

Figure 3a – *TEM Appearance of a CoCr Metal Wear Particle of the Dense, Elongated Form*

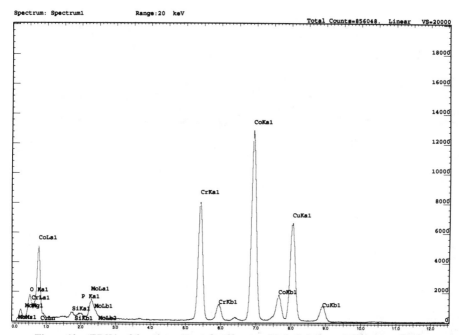

Figure 3b - *EDXA of the Above Particles Showing High Cobalt and Chromium Peaks*

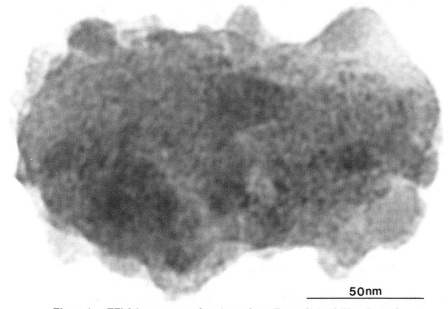

50nm

Figure 4a - *TEM Appearance of an Amorphous Type of Metal Wear Particle*

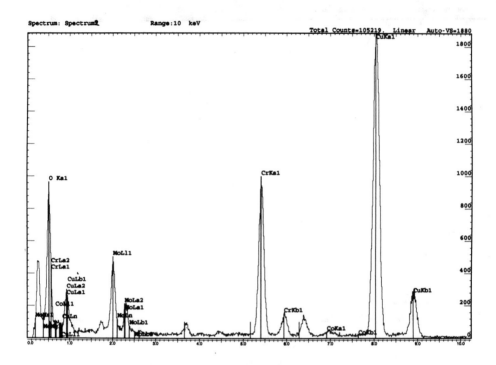

Figure 4b - *EDXA of the Above Particles Showing High Chromium and Oxygen Peaks*

Table 4 - *Results of Image Analysis of Metal Wear Particles*

	Particle Size in nanometers (nm)				
	Average	Median	Deviation	Minimum	Maximum
Metasul (n = 5)	171	115	161	16	938
SR (n = 4)	118	101	80	25	628

Discussion

The wear of all of these second-generation M-M implants was substantially lower than the 50 -100 mm^3 per year that is typical of UHMWPE acetabular cups, and most of the wear characteristics were comparable to those of the first-generation M-M implants [17], including micropitting of the bearing surfaces. Micropits have been described in retrieved McKee-Farrar THRs [19] as well as in wear tested, second-generation M-M components, although in much smaller numbers [20]. Although the mechanisms for forming the carbide depressions and the micropits in the THRs have not yet been fully determined, these did not appear to be associated with an increased wear rate.

Third body damage was noted in varying degrees on all of the components, including well-fixed cases that were retrieved at autopsy. Flattened carbides were also noted in the bearing zones in all of the THR cases. These observations suggest that an early release of carbides during the initial wear-in period is the mechanism responsible for third body wear scratching in well-fixed components [20]. The greatest amount of third body damage occurred in loose cemented implants due to PMMA particles. However, evidence of "self-polishing" of third body wear scratches, noted in the first generation components [17], was also seen in these second generation implants, including the cases that were damaged by recurring dislocation. One of those cases (THR 2) had a low surface roughness and the bearing surface was relatively free of scratches, possibly because the high activity level of the patient contributed to the rapid re-polishing of the damaged surface. In contrast, in the other dislocation case (THR 7), the surface roughness remained high and there were more scratches seen with SEM because the implants were removed shortly after the dislocation episode and there had been little time for self-polishing. In some of the SRs, the metal release in the early wear-in was exacerbated by the initial out of roundness of the components.

The periprosthetic histology and wear particle morphology of these second generation M-M THRs were consistent with the low wear of these bearings in vivo and were similar to those reported for first generation M-M THRs. For example, Willert et al. (1996) re-examined tissues from nineteen first-generation M-M THRs collected from 1970 to 1994, which were revised for loosening [21]. The histological reaction to metal wear particles was described as mild, and the predominant macrophage response was to particles of PMMA. Particles of metal were not well visualized by scanning electron microscopy of the tissues, but the authors reported that 50% of the metal particles were less than 1 µm, and 33% were less than 0.5 µm. Area measurement of the particles showed that the metal occupied only a small amount of tissue, which was consistent with the overall low wear rate of the components, approximately 5 mm^3 per year.

Wait et al. (1995) studied tissues from around 43 cemented first-generation M-M THRs obtained at revision or autopsy, using routine histology and immunohisto-chemistry. The histology was variable, but well defined granulomas were seen in twelve of these, wherein normal appearing macrophages contained metal wear particles while the surrounding tissue was relatively devoid of particles. Grey tissues were associated with tissue necrosis, with T lymphocytes in the adjacent viable tissue. CoCr particles, either clumped or individual, were found in membrane-bound vesicles within cells. The particles were elongated and often needle shaped, and most were under 20 nm.

In a previous study by our group, capsular and interface tissues from nine long and short term M-M THRs of first and second generations were studied to assess the tissue reaction around M-M prostheses [13,]. Compared to typical M-PE retrieval tissues, the extent of the inflammatory reaction to wear particles and the presence of foreign body type giant cells was much less intense in M-M tissues [13] . TEM studies showed that the metal particles were nanometer-sized, which meant that there may actually be more wear particles in the tissues around low wearing M-M THRs compared to higher wearing PE components which produce larger (micron-sized) particles.

Osteolysis was not a major cause of failure in first generation M-M THRs and has not been reported to date in modern-generation M-M patients [14]. The capsular granulomas in some of the M-M SRs in the present study may have been associated with PMMA debris caused by the loosening process even though the PMMA particles for

these cases were rated as absent or minimal. This can be explained by the fact that
PMMA particles are dissolved during the tissue processing; only the larger particles
leave histological evidence of their presence in the form of holes within, or surrounded
by, giant cells. Smaller particles within macrophages would not be apparent and would
not be reflected in the rating scores. Similarly, visible metal particles were not seen in
large numbers in the majority of cases and this was reflected in the rating even though
the tissues were light grey in color and the components had undergone measurable wear.
However, the SR tissues contained more visible metal particles than the THRs, which
probably reflected the larger amount of metal particles released during the early wear-in
of these large bearings. Initial manufacturing difficulties associated with these large
components led to a greater amount of out-of-roundness compared to the Metasul THRs
with 28 mm or 32 mm bearings, and the average wear depth was higher than in the
Metasuls *[23,24]*. Despite the formation of small granulomas within the capsules
around the SRs, osteolysis was not noted in the pre-revision radiographs and the one
component that was cut showed an intact femoral head. This is in sharp contrast to the
extensive histiocytic granulomas in the soft tissues and the large osteolytic lesions often
seen in the femoral heads treated with uncemented titanium alloy-PE surface
replacements *[25,26]*.

Thus, despite the presence of potentially large numbers of nanometer-sized metal
particles in the periprosthetic tissues, the macrophage and giant cell response to them
could be described as mild. However, immunohistochemical studies have indicated that
CoCr particles can induce the release of potentially osteolytic cytokines *[27,28]* and
potentially adverse affects including hypersensitivity *[29]* and chromosomal damage *[30]*
have been reported in the literature. Continued histopathogical studies and
measurements of in vivo wear in M-M THRs are therefore recommended.

Acknowledgment

This study was supported by Sulzer Orthopedics, Wright Medical Technologies and
by the Los Angeles Orthopaedic Hospital Foundation. We are grateful to Dr. L. Walter, and
Dr. E. Marel for providing tissue specimens from their cases.

References

[1] Wright, T. and Goodman, S., eds., *Implant Wear: The Future of Total Joint
     Replacement*, American Academy of Orthopaedic Surgeons, 1996, p. 116.
[2] McKellop, H. A., Campbell, P., Park, S. H., Schmalzried, T. P., Grigoris, P.,
     Amstutz, H. C., and Sarmiento, A., "The Origin of Submicron Polyethylene Wear
     Debris in Total Hip Arthroplasty," *Clinical Orthopaedics and Related Research*,
     Vol. 311, 1995, pp. 3-20.
[3] Shanbhag, A. S., Jacobs, J. J., Glant, T. T., Gilbert, J. L., Black, J., and Galante, J.
     O., "Composition and Morphology of Wear Debris in Failed Uncemented Total
     Hip Replacement," *Journal of Bone and Joint Surgery. British Volume*, Vol.
     76B, 1994, pp. 60-67.

[4] Campbell, P., Doorn, P., Dorey, F., and Amstutz, H. C., "Size and Morphology of UHMWPE Wear Particles from Total Hip Replacements," *Proceedings of the Institution of Mechanical Engineers. Part H, Journal of Engineering in Medicine*, 1996.

[5] Kadoya, Y., Kobayashi, A., and Ohashi, H., "Wear and Osteolysis in Total Joint Replacement," *Acta Orthopaedica Scandinavica Supplementum*, Vol. 69, 1998, pp. 1-15.

[6] Jiranek, W. A., Machado, M., Jasty, M., Jevsevar, D., Wolfe, H. J., Goldring, S. R., Goldberg, M. J., and Harris, W. H., "Production of Cytokines Around Loosened Cemented Acetabular Components. Analysis with Immunohistochemical Techniques and in Situ Hybridization," *Journal of Bone and Joint Surgery. American Volume*, Vol. 75A, 1993, pp. 863-879.

[7] Kim, K. J., Itoh, T., and Kumegawa, M., "Assessment of Osteoclast-Mediated Bone Resorption Induced by Wear Particles Using Rabbit Unfractionated Bone Cells," *Forty-first Annual Meeting of the Orthopaedic Research Society*, 1995, p. 771.

[8] Al Saffar, N. and Revell, P. A., "Interleukin-1 Production by Activated Macrophages Surrounding Loosened Orthopaedic Implants: A Potential Role in Osteolysis," *British Journal of Rheumatology*, Vol. 33, 1994, pp. 309-316.

[9] Pandey, R., Quinn, J., Joyner, C., Murray, D. W., Triffitt, J. T., and Athanasou, N. A., "Arthroplasty Implant Biomaterial Particle Associated Macrophages Differentiate into Lacunar Bone Resorbing Cells," *Annals of the Rheumatic Diseases*, Vol. 55, 1996, pp. 388-395.

[10] Murray, D. W. and Rushton, N., "Macrophages Stimulate Bone Resorption when They Phagocytose Particles.," *Journal of Bone and Joint Surgery. British Volume*, Vol. 72B, 1990, pp. 988-992.

[11] McKellop, H., Lu, B., and Wiser, H., "The Effect of Specimen Orientation and Lubricant Concentration on the Wear of Metal/Metal Hip Implants in a Wear Simulator," *Twenty-fourth Annual Meeting of the Society for Biomaterials*, 1998, p. 422.

[12] McKellop, H., Doorn, P., Chiesa, R., Park, S. H., and Amstutz, H., "Twenty-Year Retrieval Analysis of Metal-Metal Hip Prostheses," *Forty-second Annual Meeting of the Orthopaedic Research Society*, 1996, p. 456.

[13] Doorn, P. F., Mirra, J. M., Campbell, P. A., and Amstutz, H. C., "Tissue Reaction to Metal on Metal Total Hip Prostheses," *Clinical Orthopaedics and Related Research*, Vol. 329 Suppl, 1996, pp. S187-205.

[14] Amstutz, H. C., Campbell, P., McKellop, H., Schmalzreid, T. P., Gillespie, W. J., Howie, D., Jacobs, J., Medley, J., and Merritt, K., "Metal on Metal Total Hip Replacement Workshop Consensus Document," *Clinical Orthopaedics and Related Research*, Vol. 329 Suppl, 1996, pp. S297-303.

[15] Black, J., "Metal on Metal Bearings. A Practical Alternative to Metal on Polyethylene Total Joints?," *Clinical Orthopaedics and Related Research*, Vol. 329 Suppl, 1996, pp. S244-255.

[16] Merritt, K. and Brown, S. A., "Particulate Metals: Experimental Studies," *Biological, Material, and Mechanical Considerations of Joint Replacement*, Bristol-Myers Squibb. Zimmer Orthopaedic Symposium Series, Morrey, B. F.

(ed.), Raven Press, Ltd., New York, 1993, pp. 147-159.

[17] McKellop, H., Park, S.-H., Chiesa, R., Doorn, P., Lu, B., Normand, P., Grigoris, P., and Amstutz, H., "In Vivo Wear of Three Types of Metal on Metal Hip Prostheses During Two Decades of Use," *Clinical Orthopaedics and Related Research*, Vol. 329 Suppl, 1996, pp. S128-140.

[18] Doorn, P. F., Campbell, P. A., Worrall, J., Benya, P. D., McKellop, H. A., and Amstutz, H. C., "Metal Wear Particle Characterization from Metal on Metal Total Hip Replacements: Transmission Electron Microscopy Study of Periprosthetic Tissues and Isolated Particles," *Journal of Biomedical Materials Research*, Vol. 42, 1998, pp. 1013-1111.

[19] Walker, P. S., Salvati, E., and Hotzler, R. K., "The Wear on Removed McKee-Farrar Total Hip Prostheses," *Journal of Bone and Joint Surgery. American Volume*, Vol. 56A, 1974, pp. 92-100.

[20] Park, S.-H., McKellop, H., Lu, B., Chan, F., and Chiesa, R., "Wear Morphology of Metal-Metal Implants: Hip Simulator Tests Compared to Clinical Retrievals," *ASTM Symposium on Alternative Bearings*, 1997.

[21] Willert, H. G., Buchhorn, G. H. H., Göbel, D., Köster, G., Schaffner, S., Schenk, R., and Semlitsch, M., "Wear Behavior and Histopathology of Classic Cemented Metal on Metal Hip Endoprostheses," *Clinical Orthopaedics and Related Research*, Vol. 329 Suppl, 1996, pp. S160-186.

[22] Doorn, P. F., Campbell, P. A., and Amstutz, H. C., "Metal Versus Polyethylene Wear Particles in Total Hip Replacements. A Review," *Clinical Orthopaedics and Related Research*, Vol. 329 Suppl, 1996, pp. S206-216.

[23] Amstutz, H. C., Sparling, E. A., Campbell, P., McKellop, H., and Dorey, F., "Preliminary Experience with Metal/Metal Hybrid and Cementless Surface Replacements of the Hip," *Twenty-fourth Annual Meeting of the Society for Biomaterials*, 1998, p. 213.

[24] Campbell, P., McKellop, H., Lu, B., Park, S.-H., Doorn, P., Dorr, L., and Amstutz, H. C., "Clinical Wear Performance of Modern Metal-on-Metal Hip Arthroplasties," *Twenty-fourth Annual Meeting of the Society for Biomaterials*, 1998, p. 210.

[25] Amstutz, H. C., Campbell, P., Clarke, I. C., and Kossovsky, N., "Mechanism and Clinical Significance of Wear Debris-Induced Osteolysis," *Clinical Orthopaedics and Related Research*, Vol. 276, 1992, pp. 7-18.

[26] Nasser, S., Campbell, P. A., Kilgus, D., Kossovsky, N., and Amstutz, H. C., "Cementless Total Joint Arthroplasty Prostheses with Titanium-Alloy Articular Surfaces. A Human Retrieval Analysis," *Clinical Orthopaedics and Related Research*, Vol. 261, 1990, pp. 171-185.

[27] Revell, P. A., al-Saffar, N., and Kobayashi, A., "Biological Reaction to Debris in Relation to Joint Prostheses," *Proceedings of the Institution of Mechanical Engineers. Part H, Journal of Engineering in Medicine*, Vol. 211, 1997, pp. 187-197.

[28] Lee, S.-H., Brennan, F. R., Jacobs, J. J., Urban, R. M., Ragasa, D. R., and Glant, T. T., "Human Monocyte/Macrophage Response to Cobalt-Chromium Corrosion Products and Titanium Particles in Patients with Total Joint Replacements,"

Journal of Orthopaedic Research, Vol. 15, 1997, pp. 40-49.

[29] Merritt, K. and Brown, S. A., "Distribution of Cobalt Chromium Wear and Corrosion Products and Biologic Reactions," *Clinical Orthopaedics and Related Research,* Vol. 329 Suppl, 1996, pp. S233-243.

[30] Case, C. P., Langkamer, V. G., Howell, R. T., Webb, J., Standen, G., Palmer, M., Kemp, A., and Learmonth, I. D., "Preliminary Observations on Possible Premalignant Changes in Bone Marrow Adjacent to Worn Total Hip Arthroplasty Implants," *Clinical Orthopaedics and Related Research,* Vol. 329 Suppl, 1996, pp. S269-279.

Nadim J. Hallab,[1] Joshua J. Jacobs,[1] Anastasia Skipor,[1] Jonathan Black,[2] Katalin Mikecz[1] and Jorge O. Galante[1]

Serum Protein Carriers of Chromium in Patients with Cobalt-Base Alloy Total Joint Replacement Components

Reference: Hallab, N. J., Jacobs, J. J., Skipor, A., Black, J., Mikecz, K. and Galante, J.O., " Serum Protein Carriers of Chromium in Patients with Cobalt-base Alloy Total Joint Replacement Components," *Cobalt-Base Alloys for Biomedical Applications, ASTM STP 1365*, J. A. Disegi, R. L. Kennedy, and R. Pillar, Eds., American Society for Testing and Materials, West Conshohocken, PA, 1999.

Abstract: The distribution of chromium (Cr) in fractionated serum was studied from peripheral blood of patients with and without Cr-containing Cobalt-base alloy total joint replacements:1)10 patients with cobalt-chromium-base alloy prostheses; and 2) 10 age-matched controls without implants. Two molecular weight ranges were found to primarily bind Cr (at $\approx$70 kD and $\approx$140-180 kD) in patients with Cobalt-base alloy total joint replacements. This pattern of concentration-dependent metal-protein binding within molecular weight ranges that include immunoglobulins warrants further investigation.

Keywords: metal ion release, biocompatibility, protein, chromium, corrosion

Introduction

It has been well established that all metallic implants corrode *in vivo*, releasing metal into the surrounding tissue and fluid [1-5]. Elevated concentrations of metals, including Cr, Co and Ti, have been reported in the serum of patients with both well functioning and failed total joint replacements [5-7]. Up to 5-fold and 8-fold elevations in serum and urine Cr concentrations, respectively, have been found in patients with well-functioning total hip replacements [11]. The long term biologic effects of these circulating organometallic complexes are currently unknown. Identification of the ligands associated with circulating metal organometallic complexes must precede characterization of associated bioreactivity.

---

[1]Assistant Professor, Professor and Director, Section of Biomaterials, Research Associate, Associate Professor, Professor and Director Rush Arthritis and Orthopedics Institute, respectively, Rush-Presbyterian-St.-Luke's Medical Center, 1653 West Congress Parkway, Chicago, IL 60612.
[2]Principal, IMN Biomaterials, King of Prussia, PA 19406.

Studies of biologic reactivity of dissolved metal have combined free metal ions (metal salts) with proteins [8, 9] in ways which may not reproduce *in vivo* binding behavior of released soluble products. The goal of this study was to characterize the distribution of chromium within fractionated serum of patients with TJA components. We focused this study on chromium based on our previous work, which demonstrated consistant elevations in serum chromium following total hip replacement with cobalt-based alloys.

Materials and Methods

Specimen Collection

Serum samples were obtained through peripheral blood collection from patients with total joint replacements and known elevated levels of total serum Cr as identified in previous studies [4, 5, 10, 11]. Blood samples were obtained with great care to avoid contamination as previously described [6, 10]. Thirty milliliters of blood was collected, with polypropylene syringes, allowed to fully clot (20-120 min), centrifuged (1850Xg 30 min at 37°C), and the serum fraction collected. All serum was stored at minus 80°C. All collection containers and apparatus were triple acid-washed with Ultrex-grade chemicals (Baker, Chicago, IL) or verified to be metal contamination free by flushing with double deionized water followed by measurement of any residual trace metals with atomic absorption spectroscopy. All manipulations of the specimens were carried out using class-100 gloves (Oak Technical, Ravenna, IL) and in a class-100 environment within Sterilguard Hoods (Baker, Sanford, ME).

Patient Groups

Two groups of patients were studied. These two groups of patients were selected to represent, (1) people without implants who had normal levels of serum chromium and (2) patients with a variety total hip replacements (THR) who had previously determined high levels of circulating from implant degradation. THR patients were selected with serum metal Cr concentrations >1 parts per billion (ppb). Group A contained 10 age matched controls with no implants and no systemic disease (5 Female, 5 Male, Ages 32-75, Average Age 61). The average Cr content of Group A controls was 0.07 ppb, when measured in serum. Group B contained 10 patients with Co-Cr base alloy implants and known elevated levels of serum Cr (7 Female, 3 Male, Ages 26-73, Average Age 52). The average Cr content of Group B patients was 2.27 ppb Cr (32-fold that of the control group A). Subjects within both groups were reported to be otherwise healthy and symptom free.

Serum Fractionation and Total Protein Measurement

Serum samples were fractionated based on molecular size using fast-protein-liquid-chromatography (FPLC). FPLC was selected for serum protein separation in order to minimize metal contamination of serum samples by processing samples in a clean all-polymer environment, which was verified as contamination free through preliminary testing of eluent. The FPLC configuration was constructed to maximize the amount of serum sample available for fractionation and metal content analysis. A P-500 Pump (Pharmacia, Piscataway, NJ) was used to pump a 1 mL sample of serum through two 24 mL Superdex™ 200 columns (Pharmacia, Piscataway, NJ). These columns were mounted in series. Double deionized water (>18 MOhms) from a Milli-Q water system (Millipore, Bedford, MA) was used as a metal ion contamination-free eluent at flow rate of 5 mL/hr. Other possible eluent buffers such as Ultrex grade KPO_4 and NaOH were found to contain metal within the parts per million range precluding their use as buffer solutions. Detection of protein exiting the columns was carried out using a UV-1 cell ultra-violet absorption monitor, (Pharmacia, Piscataway, NJ). High molecular weight standard protein calibrants (Pharmacia, Piscataway, NJ) were used both with FPLC and in conjunction with polyacrylamide gel electrophoresis to determine the approximate molecular weight of prominent proteins within each FPLC fraction. It is possible that metal binding and other protein-protein interactions skew the approximations of molecular weight ranges. Therefore it must be clear that molecular weight ranges associated with FPLC fractions are best approximations given the current technique. Post column reconcentration of samples was avoided to reduce contamination through less sample processing. Practical limitations regarding the time necessary to analyze serum fractions using Graphite furnace Zeeman atomic absorption spectroscopy (GFZ-AAS) restricted the number of fractions to 11.

Assessment of total protein was conducted on serum samples pre and post fractionation using Pierce Protein reagents. Total protein assays were conducted using 10 μl of sample with 200μl of substrate in 96-well plates, with serial dilutions of bovine serum albumin used as standard calibrants.

Metal Content Determination

Metal content within serum fractions was analyzed using a graphite furnace Zeeman atomic adsorption spectrophotometer (GFZ-AAS) (Perkin-Elmer, Norwalk, CT) with an HGA 600 heated graphite atomizer and an AS 60 autosampler (Perkin-Elmer, Norwalk, CT) using methods previously described [10]. The method detection limit of metal content was 0.6 ng/mL for Cr in diluted serum fractions post fractionation by FPLC (detection limits in undiluted serum are 0.03 ng/mL for Cr and 0.6 ng/mL following FPLC).

Statistical Analysis

By convention, Cr concentrations below detection limits are expressed as one-half the method detection limit concentration to provide a relative scale for graphical comparison. Intergroup comparisons, independent of these means, were made using Kruskall-Wallis non-parametric analysis of variance since groups had left-censored data (normal values less than the method detection limit). The Wilcoxon-Mann-Whitney test was then used if the Kruskall-Wallis test revealed significant differences at the $p < 0.05$ level.

Results

A typical chromatogram produced by FPLC fractionation is shown in Figure 1. Eleven molecular weight fractions were collected. Chromatograms (serum protein profiles) for all groups showed the same basic profile of molecular weight vs. protein amount.

Similarity of protein distribution in all fractions of Groups A and B can be seen in Table 1 which shows means of protein content for each of the fractions for each patient group. These values are listed in Table 1 where the recovery of protein after fractionation is shown compared to that directly measured from serum before fractionation. Both groups show recovery values greater than 91%.

However, despite similarity between intergroup total protein, there were significant ($p < 0.05$) differences between levels of metal detected in certain serum fractions of Group A (controls) and those of Group B (patients with implants). These results are listed in Table 2. The mean Cr levels detected in the fractionated serum of Group A are compared to those of Group B patients (Figure 2). In this figure Cr exhibits a bimodal binding pattern to serum proteins, binding both midrange molecular weight protein(s) at 35-70 kD and higher molecular weight protein(s) in the 140-180 kD range.

Serum metal was not detected in any chromatography fractions associated with either low molecular weight serum proteins or in that of eluent preceding serum proteins we conclude that little, if any, metal exists as unbound ionic species. Additionally, pressurized filtration of serum containing metal through 1 kD filters was used to verify that metal was not bound by low molecular weight serum proteins were present as colloidally suspended hydroxides.

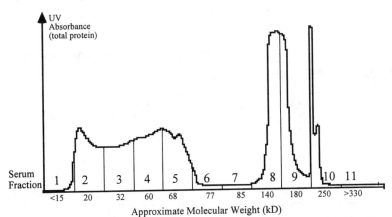

Figure 1. Typical FPLC Chromatogram, showing 11 serum protein fractions (and associated approximate molecular weights) collected for metal ion analysis.

Table 1-*Average amount of protein (mg/mL) within each of the eleven fractions for each patient group.*

Protein(mg/mL)	Total Protein (mg/mL)											Total Serum	
Serum Fraction (~kD)	1 <15	2 20	3 32	4 60	5 68	6 77	7 85	8 140	9 180	10 220	11 >330	Recovered (post-FPLC)	Direct (pre-FPLC)
Group A n=10	*	40.7	59.4	101.1	100.0	28.9	*	7.2	63.3	9.9	*	46.2	50.6
Group B n=10	*	39.1	70.4	127.9	115.8	*	*	3.1	72.3	3.7	*	49	48.8

*Below method detection limit (<4 mg/mL protein adjusted for dilution, <0.2 mg/mL pre-FPLC)

Therefore, the preponderance of Cr detected in the serum of patients with elevated levels of circulating metal was conjugated with mid- and high-range serum proteins.

The bimodal distribution of Cr is apparent in Groups A and B serum fractions; however, this binding pattern is subtler in Group A, with no statistically significant increases in Cr concentration upon intergroup comparison. Both these Cr binding ranges show statistically significant ($p<0.05$) increases of approximately 4 to 7 times the average Cr content of controls.

Table 2-*Averaged Cr concentrations within Groups A and B for serum fractions 1 through 11*

Serum Fraction	1	2	3	4	5	6	7	8	9	10	11
Molecular Weight (~kD)	<15	20	32	60	68	77	85	140	180	220	>330
Group A, Controls (n=10)											
Cr (ppb)	*	0.66	1.11	1.11	1.16	0.67	0.51	0.87	0.68	0.54	0.34
Group B, Patients with Co-Cr base Implants, (n=10)											
Cr (ppb)	*	1.4	4.0	7.4	7.9	1.3	1.1	2.2	8.3	2.8	*

* Below detection limit

Discussion

The presence of elevated concentrations of chromium in the serum of patients with total joint replacements has been described in multiple investigations [11-13, 14]. It is unclear whether the circulating metal is preferentially bound by relatively ion-specific protein binding sites, non-specifically bound by bulk serum proteins such as albumin, or whether the metal exists in a non-ionic form. The strengths of metal-protein bonds and their consequent bioreactivity are as yet largely uncharacterized. Metals vary in their relative affinities for different biological molecules with the physiochemical properties of each metal directly affecting how it interacts.

Metal-protein binding associated with biomaterial degradation has been previously investigated to some extent *in vitro* [14-17] but *in vivo* human investigations have not been previously conducted on patients with metallic implants. Previous investigations do not reach a consensus as to which serum proteins primarily bind metals such as Cr [14-17]. However, serum proteins in molecular weight ranges, which include albumin, were implicated as binding some degree of metal in all these studies [14-18].

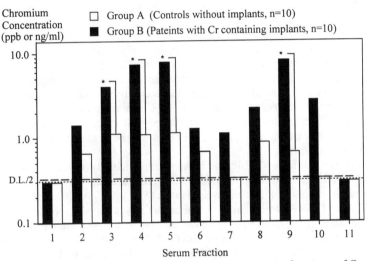

Figure II. Comparison of the average Cr content in serum fractions of Group A controls and Group B patients with Co-base implants and elevated levels of serum Cr. Note: Detection limit (D.L.) denoted by dotted line and (*) indicates statistical significance at p<0.05.

The disagreement between these reports lies in the identification of which serum proteins beside albumin bind metal and which serum protein(s) dominate(s) metal binding. Merritt et al. [17] found most Ni bound to albumin (<70 kD) whereas Woodman et al. [16] found a relatively equal distribution between an albumin range (≈66 kD) and a high gamma globulin range (≈140-820 kD).

Recognizing that metal release from implants always occurs to some extent, we hypothesized that: 1) essentially all soluble metals released from implants are bound to proteins and that 2) specific metal-protein binding is more prevalent than nonspecific binding of metal to all serum protein. The comparatively high Cr metal content found here within specific molecular weight ranges of protein support these hypotheses.

However, this does not rule out other *in vivo* binding sites for released metal, such as bone mineral, other extracellular tissue and local and remote cells.

Within the serum of Group B patients, the bimodal pattern of Cr-protein binding was primarily in two molecular weight ranges (≈60-77 kD and ≈180 kD). Similar serum proteins were reported by Woodman et al. [16] to bind metal in rat serum. An investigation by Merritt et al. [17] found Ni and Cr bound to rabbit serum protein ranges associated with albumin (<70 kD) and gamma globulins (≈200 kD), although binding was reported to be primarily to albumin. It may have been incorrect of some previous reports to assume nonspecific metal-protein binding dominated *in vivo*, implying albumin dominantly bound metal because of its prevalency in serum and interstitial tissue [19, 20]. Others such as Borguet et al. [18] found that ≈80% of radiolabeled ionic Cr added to the serum of dialysis patients (separated using anion exchange chromatography) was bound to transferrin (≈77 kD) and the remaining ≈20% was bound to albumin.

Little is known of the bioreactivity of *in vivo* metal-protein complexes detected in the serum of patients with total joint arthroplasties formed from elevated levels of serum metal since this is the first report identifying metal-protein complexes in patients with total joint arthroplasties. The affinity and avidity of particular ligand binding to Cr released from implants, can determine resultant bioreactivity by controlling the ease with which cellular internalization of metal may occur [21]. There may be greater Cr bioreactivity at the intracellular level when complexed with easily cell-internalized metalloproteins such as transferrin. These differences may be important in evaluation of different types of implants.

Protein within the range of ≈140-250 kD may be particularly relevant to the evaluation of possible immunologic or toxic interactions due to the prevalence of immunoglobulins within this molecular weight range (e.g., IgG, IgA, IgD, IgE - immunoglobulins, Table 3). Speculation of immunoactivity is supported by investigations such as that by Yang et al. [20] who have shown binding of humoral antibodies (IgG, IgE, IgM and IgA, Table 3) to metal complexed with Glutathione (GSH). Efforts are currently underway to more specifically identify which proteins are associated with binding metal within the molecular weight fractions identified in this investigation.

Conclusions

In this investigation we have identified two ranges of human serum proteins associated with the *in vivo* binding of Cr from implant degradation. The toxicologic significance of chronic low level exposure to Cr from metallic implant degradation is not well documented [7, 10]. Prerequisite to the prediction of potential toxicity related to chronic low level exposure of metal released from implants, is a clear understanding of which proteins are involved and how strongly they bind metal. However, the results of this investigation are the first to show this metal binding phenomenon in human patients with metallic implants. As the longer term results of implant performance become available, and as newer prosthetic designs are introduced that are intended for use in younger, more active individuals, there is an increasing need for complete characterization of the bioavailability and bioreactivity of circulating metal from implant degradation.

Table 3 - *Important serum proteins associated with each FPLC serum fraction* [22].

Serum Fractions and prominant serum proteins associated with each fraction	Concentration in Serum gm/L)(mg/100 mL)	Approximate Molecular Weight (kD)
Fraction 1		<15
Fraction 2		20
β2-Microglobulin	0.2	12
Lysozyme	7-20	15
Rentinol-binding protein	4	21
Fraction 3		32
B2-Glycoprotein III	5-15	35
Thyroxin binding Glubulin	1-2	36
a&B-glycoprotein	15-140	40-44
Gc-Globulin	20-55	51
α-antitrypsin	200	54
Fraction 4		60
Anti-thrombin III	17-30	65
Albumin	3500-5500	66
Fraction 5		68
Albumin	3500-5500	66
α1-Antichymotrypsin	30-60	68
Fraction 6		77
Transferrin	200-400	77
C9-complement factor	0.1-1	79
Fraction 7		85
Haptoglobin (2-1 type)	160-300	80
Hemopexin	50-115	80
Plasminogen	10-30	81
C1s-complement factor	11	86
C6-complement factor	7.5	95
Haptoglobin (1-1 type)	100-220	100
Fraction 8		140
C7-complement factor	5.5	110
C2-complement factor	2.5	117
Haptoglobin (2-2 type)	120-160	120-160
IgG	800-1800	150
C8-complement factor	50	153
IgA	90-450	160
Fraction 9		180
α1-Lipoprotein	360	180
C3-complement factor	160	180-350
C5-complement factor	8	180
IgG	800-1700	150
IgA	100-400	150
C8-complement factor	50	163
IgD	20	170
IgE	0.001	190
Fraction 10		250
C4-complement factor	20-50	230
Fraction 11		>330
C1q-complement factor	10-25	400
α2-macroglobulin	200	820
IgM	5-200	900
β lipoproteins	400	>2,000
β-Liproteins, LP-B	220-740	2,400

References

[1]   Ferguson-Jr., A.B., Laing, P.G. and Hodge, E.S., "The ionization of metal implants in living tissue," *Journal of Bone and Joint Surgery*, Vol. 42-A, 1960, pp 77-90.

[2]   Agins, H., Alcock, N.W., Bansal, M., Salvati, E.A., Wilson, P.D., Pellicci, P.M. and Bullough, P.G., "Metallic wear in failed titanium-alloy total hip replacements. A histological and quantitative analysis," *Journal of Bone and Joint Surgery*, Vol. 70-A, 1988, pp 347-356.

[3]   Black, J., Sherk, H., Bonini, J., Rostoker, W.R., Schajowicz, F. and Galante, J.O., "Metallosis associated with a Stable titanium-alloy femoral component in total hip replacement," *Journal of Bone and Joint Surgery*, Vol. 72-A, 1990, pp 126-130.

[4]   Jacobs, J.J., Skipor, A.K., Black, J.M., L., Urban, R.M. and Galante, J.O., "Metal Release in Patients With Loose Titanium Alloy Total Hip Replacements," *Transactions of the Fourth World Biomaterials Conference*, Berlin, 1992, pp 266.

[5]   Jacobs, J.J., Skipor, A.K., Doorn, P.F., Campbell, P., Schmalzried, T.P., Black, J. and Amstutz, H.C., "Cobalt and chromium concentrations in patients with metal on metal total hip replacements," *Clinical Orthopaedics and Related Research*, Vol. S329, 1996, pp S256-S263.

[6]   Jacobs, J.J., Skipor, A.K., Black, J., Urban, R.M. and Galante, J.O., "Metal Release and Excretion in Patients with Titanium Base Alloy Total Hip Replacement Components," *Journal of Bone and Joint Surgery*, Vol. 73A, 1991, pp 1475-1486.

[7]   Jacobs, J.J., Skipor, A.K., Black, J., Hastings, M.C., Schavocky, J., Urban, R.M. and Galante, J.O., "Metal Release and Excretion from Cementless Titanium Total Knee Replacement.," *Trans ORS*, 16, 1991, pp 558.

[8]   Merritt, K. and Brown, S., "Tissue reaction and metal sensitivity," *Acta Orthopaedica Scandinavia*, Vol. 51, 1980, pp 403-4111.

[9]   Brown, G.C., Lockshin, M.D., Salvati, E.A. and Bullough, P.G., "Sensitivity to metal as a possible cause of sterile loosening after cobalt-chromium total hip-replacement arthroplasty," *The Journal of Bone and Joint Surgery*, Vol. 59-A, 1977, pp 164-168.

[10] Jacobs, J.J., Silverton, C., Hallab, N.J., Skipor, A.K., Patterson, L.M., Black, J., and Galante, G.O., "Metal release and excretion from cementless titanium alloy total knee replacements," *Clinical Orthopeadics and Related Research*, Vol. 358, 1999, pp 173-180.

[11] Jacobs, J.J., Skipor, A.K., Patterson, L.M., Hallab, N.J., Paprosky, W.G., Black, J. and Galante, J.O., "A prospective, controlled, longitudinal study of metal release in patients undergoing primary total hip arthroplasty," *Jornal of Bone and Joint Surgery*, Vol. 80-A, 1998, pp 1447-1458.

[12] Sunderman, W., Hopfer, S.M., Swift, T., Rezuke, W.N., Ziebka, L., Highman, P., Edwards, B., Folicik, M. and Gossling, H.R., "Cobalt, chromium, and nickel concentrations in body fluids of patients with porous-coated knee or hip prosthesis," *Journal of Orthopedic Research*, Vol. 7, 1989, pp 307-315.

[13] Jacobs, J., Urban, R.M., Gilbert, J.L., Skipor, A., Black, J., Jasty, M.J. and Galante, J.O., "Local and distant products from modularity," *Clinical Orthopedics*, Vol. 319, 1995, pp 94-105.

[14] Black, J., Maitin, E.C., Gelman, H. and Morris, D.M., "Serum concentrations of chromium, cobalt, and nickel after total hip replacement: a six month study," *Biomaterials*, Vol. 4, 1983, pp 160-164.

[15] Yang, J. and Black, J., "Competitive binding of chromium cobalt and nickel to serum proteins," *Biomaterials*, Vol. 15, 1994, pp 262-268.

[16] Woodman, J.L., Black, J. and Jiminez, S.A., "Isolation of serum protein organometallic corrosion products from 316L and HS-21 in vitro and in vivo," *Journal of Biomedical Materials Research*, Vol. 18, 1984, pp 99-114.

[17] Merritt, K., "Role of medical materials, both in implant and surface applications, in immune response and in resistance to infection," Vol. 1984, pp

[18] Borguet, F., Cornelis, R., Delanghe, J., Lambert, M. and Lameire, N., "Study of the chromium binding in plasma of patients on continuous ambulatory peritoneal dialysis," *Clinica Chimica Acta*, Vol. 238, 1995, pp 71-84.

[19] Merritt, K., "The binding of metal salts and corrosion products to cells and proteins in vitro," *Journal of Biomedical Materials Research*, Vol. 18, 1984, pp 1005-1015.

[20] Yang, J. and Merritt, K., "Detection of antibodies against corrosion products in patients after Co-Cr total joint replacements," *Jornal of Biomedical Materials Research*, Vol. 28, 1994, pp 1249-1258.

[21] Foulkes, E., "Metal disposition: An analysis of underlying mechanisms," *Metal Toxicology*. R. A. Goyer, C. D. Klaasen and M. P. Waalkes, Ed., Academic Press, New York, 1995, pp 3-29.

[22] Ritzman, S.E., Daniels, J.C., "Serum Protein abnormalities Diagnostic and Clinical Aspects," Boston, Little, Brown and Co., 1975, pp. 528-535.

Author Index

Subject Index

223

T

Tensile strength, 71
    test, 47
Tensile test, 3
Tetrahedral amorphous
    coatings, 169
Titanium
    alloy, 71
    nitride, 169

V

Vacuum induction melt, 11
Vitallium, 89

W

Wear performance, 194
Wear resistance, 11, 125, 135,
    145
Wear resistant coatings, 32, 169
Wear testing, 135, 145, 156

Wear, ultrahigh molecular
    weight polyethylene, 32
Wrought-annealed materials,
    108
Wrought cobalt-chromium-
    molybdenum, 11, 62, 98,
    125

X

X-ray diffraction, 108

Y

Yield strength, 3, 108

Z

Zirconium nitride, 169